Yamaha RD 400 Twin Owners Workshop Manual

by Mansur Darlington

Models covered:

Yamaha RD400 398 cc Introduced into UK 1976, USA 1975
Yamaha RD400E 398 cc Introduced into UK 1978

ISBN 978 0 85696 548 7

Printed in the UK   *(333 - 706)*

ABCDE
FGHIJ
KL

2

Haynes Publishing Group
Sparkford Nr Yeovil
Somerset BA22 7JJ England

Haynes Publications, Inc
859 Lawrence Drive
Newbury Park
California 91320 USA

Acknowledgements

Our grateful thanks are due to Mitsui Machinery Sales (UK) Limited, who gave permission to use the line drawings used throughout this manual and the cover photograph, and also to Mr. Brian Hamilton-Farey, Service Manager of that company, who provided much useful technical information.

Our thanks are also due to Jim Patch of Yeovil Motor Cycle Services, who supplied the Yamaha RD400C model featured in this manual.

Brian Horsfall gave considerable assistance with the strip-down and rebuilding and devised the ingenious methods for overcoming the lack of service tools. Les Brazier arranged and took the photographs that accompany the text. Jeff Clew edited the text.

Finally, we would also like to thank the Avon Rubber Company, who kindly supplied illustrations and advice about tyre fitting; NGK Spark Plugs for information on plug maintenance and Renold Limited for advice on chain care and renewal.

About this manual

The author of this manual has the conviction that the only way in which a meaningful and easy to follow text can be written is first to do the work himself, under conditions similar to those found in the average household. As a result, the hands seen in the photographs are those of the author. Unless specially mentioned, and therefore considered essential, Yamaha special service tools have not been used. There is invariably some alternative means of loosening or removing a vital component when service tools are not available, but risk of damage should always be avoided.

Each of the six Chapters is divided into numbered sections. Within these sections are numbered paragraphs. Cross reference throughout the manual is quite straightforward and logical. When reference is made 'See Section 6.10' it means Section 6, paragraph 10 in the same Chapter. If another Chapter were meant, the reference would read, for example, 'See Chapter 2, Section 6.10'. All the photographs are captioned with a section/paragraph number to which they refer and are relevant to the Chapter text adjacent.

Figures (usually line illustrations) appear in a logical but numerical order, within a given Chapter. Fig. 1.1 therefore refers to the first figure in Chapter 1.

Left-hand and right-hand descriptions of the machines and their components refer to the left and right of a given machine when the rider is seated normally.

Motorcycle manufacturers continually make changes to specifications and recommendations, and these, when notified, are incorporated into our manuals at the earliest opportunity.

Whilst every care is taken to ensure that the information in this manual is correct no liability can be accepted by the authors or publishers for loss, damage or injury caused by any errors in or omissions from the information given.

Contents

1976 Yamaha RD 400C model

Introduction to the Yamaha RD400 Twin

Although the history of Yamaha can be traced back to the year 1887, when a then very small company commenced manufacture of reed organs, it was not until 1954 that the company became interested in motor cycles. As can be imagined, the problems of marketing a motor cycle against a background of musical instruments manufacture were considerable. Some local racing successes and the use of hitherto unknown bright colour schemes helped achieve the desired results and in July 1955 the Yamaha Motor Company was established as a separate entity, employing a work force of less than 100 and turning out some 300 machines a month.

Competition successes continued and with the advent of tasteful styling that followed Italian trends, Yamaha became established as one of the world's leading motor cycle manufacturers. Part of this success story is the impressive list of Yamaha 'firsts' - a whole string of innovations that include electric starting, pressed steel frame, torque induction and 6 and 8 port engines. There is also the "Autolube" system of lubrication, in which the engine-driven oil pump is linked to the twist grip throttle, so that lubrication requirements are always in step with engine demands.

Since 1964, Yamaha has gained the World Championship on numerous occasions, in both the 125 cc and 250 cc classes. Indeed, Yamaha has dominated the lightweight classes in international road racing events to such an extent in recent years that several race promoters are now instituting a special type of event in their programme from which Yamaha machines are barred! Most of the racing successes have been achieved with twin cylinder two-strokes and the practical experience gained has been applied to the road going versions.

The RD400 model was a modified and enlarged replacement of the earlier RD350, itself a decendant of the YR5.

Ordering spare parts

When ordering spare parts for the Yamaha 400 cc twins, it is advisable to deal direct with an official Yamaha agent, who will be able to supply many of the items required ex-stock. Although parts can be ordered from Yamaha direct, it is preferable to route the order via a local agent even if the parts are not available from stock. He is in a better position to specify exactly the parts required and to identify the relevant spare part numbers so that there is less chance of the wrong part being supplied by the manufacturer due to a vague or incomplete description.

When ordering spares, always quote the frame and engine numbers in full, together with any prefixes or suffixes in the form of letters. The frame number is found stamped on the right-hand side of the steering head, in line with the forks. The engine number is stamped on the left-hand side of the upper crankcase, immediately below the left-hand carburettor.

Use only parts of genuine Yamaha manufacture. A few pattern parts are available, sometimes at cheaper prices, but there is no guarantee that they will give such good service as the originals they replace. Retain any worn or broken parts until the replacements have been obtained; they are sometimes needed as a pattern to help identify the correct replacement when design changes have been made during a production run.

Some of the more expendable parts such as spark plugs, bulbs, tyres, oils and greases etc., can be obtained from accessory shops and motor factors, who have convenient opening hours, charge lower prices and can often be found not far from home. It is also possible to obtain parts on a Mail Order basis from a number of specialists who advertise regularly in the motor cycle magazines.

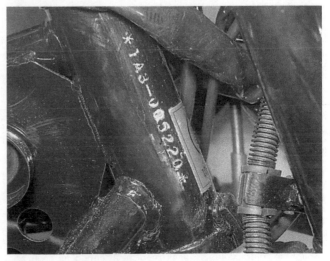

Location of engine number

Location of frame number

Safety first!

Professional motor mechanics are trained in safe working procedures. However enthusiastic you may be about getting on with the job in hand, do take the time to ensure that your safety is not put at risk. A moment's lack of attention can result in an accident, as can failure to observe certain elementary precautions.

There will always be new ways of having accidents, and the following points do not pretend to be a comprehensive list of all dangers; they are intended rather to make you aware of the risks and to encourage a safety-conscious approach to all work you carry out on your vehicle.

Essential DOs and DON'Ts

DON'T start the engine without first ascertaining that the transmission is in neutral.

DON'T suddenly remove the filler cap from a hot cooling system – cover it with a cloth and release the pressure gradually first, or you may get scalded by escaping coolant.

DON'T attempt to drain oil until you are sure it has cooled sufficiently to avoid scalding you.

DON'T grasp any part of the engine, exhaust or silencer without first ascertaining that it is sufficiently cool to avoid burning you.

DON'T allow brake fluid or antifreeze to contact the machine's paintwork or plastic components.

DON'T syphon toxic liquids such as fuel, brake fluid or antifreeze by mouth, or allow them to remain on your skin.

DON'T inhale dust – it may be injurious to health (see *Asbestos* heading).

DON'T allow any spilt oil or grease to remain on the floor – wipe it up straight away, before someone slips on it.

DON'T use ill-fitting spanners or other tools which may slip and cause injury.

DON'T attempt to lift a heavy component which may be beyond your capability – get assistance.

DON'T rush to finish a job, or take unverified short cuts.

DON'T allow children or animals in or around an unattended vehicle.

DON'T inflate a tyre to a pressure above the recommended maximum. Apart from overstressing the carcase and wheel rim, in extreme cases the tyre may blow off forcibly.

DO ensure that the machine is supported securely at all times. This is especially important when the machine is blocked up to aid wheel or fork removal.

DO take care when attempting to slacken a stubborn nut or bolt. It is generally better to pull on a spanner, rather than push, so that if slippage occurs you fall away from the machine rather than on to it.

DO wear eye protection when using power tools such as drill, sander, bench grinder etc.

DO use a barrier cream on your hands prior to undertaking dirty jobs – it will protect your skin from infection as well as making the dirt easier to remove afterwards; but make sure your hands aren't left slippery. Note that long-term contact with used engine oil can be a health hazard.

DO keep loose clothing (cuffs, tie etc) and long hair well out of the way of moving mechanical parts.

DO remove rings, wristwatch etc, before working on the vehicle – especially the electrical system.

DO keep your work area tidy – it is only too easy to fall over articles left lying around.

DO exercise caution when compressing springs for removal or installation. Ensure that the tension is applied and released in a controlled manner, using suitable tools which preclude the possibility of the spring escaping violently.

DO ensure that any lifting tackle used has a safe working load rating adequate for the job.

DO get someone to check periodically that all is well, when working alone on the vehicle.

DO carry out work in a logical sequence and check that everything is correctly assembled and tightened afterwards.

DO remember that your vehicle's safety affects that of yourself and others. If in doubt on any point, get specialist advice.

IF, in spite of following these precautions, you are unfortunate enough to injure yourself, seek medical attention as soon as possible.

Asbestos

Certain friction, insulating, sealing, and other products – such as brake linings, clutch linings, gaskets, etc – contain asbestos. *Extreme care must be taken to avoid inhalation of dust from such products since it is hazardous to health.* If in doubt, assume that they *do* contain asbestos.

Fire

Remember at all times that petrol (gasoline) is highly flammable. Never smoke, or have any kind of naked flame around, when working on the vehicle. But the risk does not end there – a spark caused by an electrical short-circuit, by two metal surfaces contacting each other, by careless use of tools, or even by static electricity built up in your body under certain conditions, can ignite petrol vapour, which in a confined space is highly explosive.

Always disconnect the battery earth (ground) terminal before working on any part of the fuel or electrical system, and never risk spilling fuel on to a hot engine or exhaust.

It is recommended that a fire extinguisher of a type suitable for fuel and electrical fires is kept handy in the garage or workplace at all times. Never try to extinguish a fuel or electrical fire with water.

Note: *Any reference to a 'torch' appearing in this manual should always be taken to mean a hand-held battery-operated electric lamp or flashlight. It does* **not** *mean a welding/gas torch or blowlamp.*

Fumes

Certain fumes are highly toxic and can quickly cause unconsciousness and even death if inhaled to any extent. Petrol (gasoline) vapour comes into this category, as do the vapours from certain solvents such as trichloroethylene. Any draining or pouring of such volatile fluids should be done in a well ventilated area.

When using cleaning fluids and solvents, read the instructions carefully. Never use materials from unmarked containers – they may give off poisonous vapours.

Never run the engine of a motor vehicle in an enclosed space such as a garage. Exhaust fumes contain carbon monoxide which is extremely poisonous; if you need to run the engine, always do so in the open air or at least have the rear of the vehicle outside the workplace.

The battery

Never cause a spark, or allow a naked light, near the vehicle's battery. It will normally be giving off a certain amount of hydrogen gas, which is highly explosive.

Always disconnect the battery earth (ground) terminal before working on the fuel or electrical systems.

If possible, loosen the filler plugs or cover when charging the battery from an external source. Do not charge at an excessive rate or the battery may burst.

Take care when topping up and when carrying the battery. The acid electrolyte, even when diluted, is very corrosive and should not be allowed to contact the eyes or skin.

If you ever need to prepare electrolyte yourself, always add the acid slowly to the water, and never the other way round. Protect against splashes by wearing rubber gloves and goggles.

Mains electricity

When using an electric power tool, inspection light etc which works from the mains, always ensure that the appliance is correctly connected to its plug and that, where necessary, it is properly earthed (grounded). Do not use such appliances in damp conditions and, again, beware of creating a spark or applying excessive heat in the vicinity of fuel or fuel vapour.

Ignition HT voltage

A severe electric shock can result from touching certain parts of the ignition system, such as the HT leads, when the engine is running or being cranked, particularly if components are damp or the insulation is defective. Where an electronic ignition system is fitted, the HT voltage is much higher and could prove fatal.

Routine maintenance

Periodic routine maintenance is a continuous process that commences immediately the machine is used. It must be carried out at specified mileage recordings, or on a calendar basis if the machine is not used frequently, whichever is the sooner. Maintenance should be regarded as an insurance policy, to help keep the machine in the peak of condition and to ensure long, trouble-free service. It has the additional benefit of giving early warning of any faults that may develop and will act as a regular safety check, to the obvious advantage of both rider and machine alike.

The various maintenance tasks are described under their respective mileage and calendar headings. Accompanying diagrams are provided, where necessary. It should be remembered that the interval between the various maintenance tasks serves only as a guide. As the machine gets older or is used under particularly adverse conditions, it would be advisable to reduce the period between each check.

For ease of reference each service operation is described in detail under the relevant heading. However, if further general information is required, it can be found within the manual under the pertinent section heading in the relevant Chapter.

In order that the routine maintenance tasks are carried out with as much ease as possible, it is essential that a good selection of general workshop tools are available.

Included in the kit must be a range of metric ring or combination spanners, a selection of crosshead screwdrivers and at least one pair of circlip pliers.

Additionally, owing to the extreme tightness of most casing screws on Japanese machines, an impact screwdriver, together with a choice of large or small crosshead screw bits, is absolutely indispensable. This is particularly so if the engine has not been dismantled since leaving the factory.

Weekly or every 200 miles

1 Topping up engine oil

Check that there is sufficient lubricant in the reservoir which feeds the mechanical engine oil pump. To ascertain the level, lift the dual seat and unscrew and withdraw the dipstick. Under no circumstances should the oil level be allowed to drop below the minimum mark. It is wise to cultivate the habit of checking the oil level every time the machine is refilled with petrol. In this way the level should not become critically low.

A warning light is provided that will illuminate when the oil will soon need replenishing. The correct oil for use in the tank is SAE 30 two-stroke engine oil, such as Castrol TT Two-stroke Oil.

2 Tyre pressures

Check the tyre pressures with a pressure gauge that is known to be accurate. Always check the pressure when the tyres are cold. If the machine has travelled a number of miles, the tyres will have become hotter and consequently the pressure will have increased. A false reading will therefore result.

Tyre pressures	Solo	two-up *
Front	26 psi (1.8 kg/cm^2)	28 psi (2.0 kg/cm^2)
Rear	28 psi (2.0 kg/cm^2)	33 psi (2.3 kg/cm^2)

These pressures should be used when the machine is used for continuous high speed riding, either solo or with a pillion passenger.

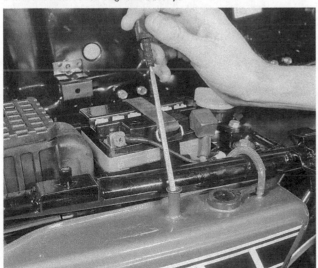

Check oil level by means of dipstick

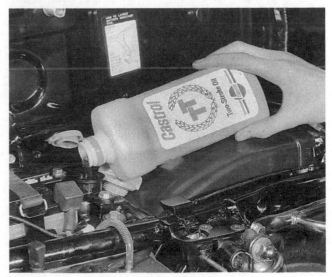

Always use the correct two-stroke oil

3 Hydraulic fluid level

Check the level of the hydraulic fluid in the front brake master cylinder reservoir, on the handlebars, and also the rear brake reservoir, behind the right-hand frame side cover. The level in both reservoirs should lie between the upper and lower level marks. During normal service, it is unlikely that the hydraulic fluid level will fall dramatically, unless a leak has developed in the system. If this occurs, the fault should be remedied AT ONCE. The level will fall slowly as the brake linings wear and the fluid deficiency should be corrected, when required. Always use an hydraulic fluid of DOT 3 or SAE J1703 specification, and if possible do not mix different types of fluid, even if the specifications appear the same. This will preclude the possibility of two incompatible fluids being mixed and the resultant chemical reactions damaging the seals.

If the level in either reservoir has been allowed to fall below the specified limit, and air has entered the system, the brake in question must be bled, as described in Chapter 5, Section 8.

4 Transmission oil

Unscrew the filler cap on the right-hand engine cover and check the transmission oil level by means of the integral dipstick. Replenish, if necessary, with SAE 10W/30 engine oil.

5 Battery electrolyte level

1 A Furukawa or Yuasa battery is fitted as standard. This battery is a lead-acid type and has a capacity of 5.5 amp hours.
2 The transparent plastic case of the battery permits the upper and lower levels of the electrolyte to be observed when the battery is lifted from its housing below the dual seat. Maintenance is normally limited to keeping the electrolyte level between the prescribed upper and lower limits and by making sure that the vent pipe is not blocked. The lead plates and their separators can be seen through the transparent case, a further guide to the general condition of the battery.
3 Unless acid is spilt, as may occur if the machine falls over, the electrolyte should always be topped up with distilled water, to restore the correct level. If acid is spilt on any part of the machine, it should be neutralised with an alkali such as washing soda and washed away with plenty of water, otherwise serious corrosion will occur. Top up with sulphuric acid of the correct specific gravity (1.260 - 1.280) only when spillage has occurred. Check that the vent pipe is well clear of the frame tubes or any of the other cycle parts, for obvious reasons.

6 Control cable lubrication

Apply a few drops of motor oil to the exposed inner portion

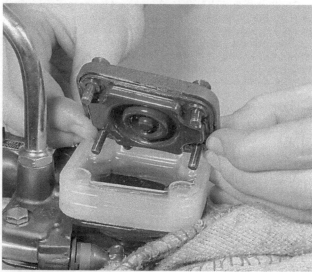

Front brake master cylinder cover retained by four screws

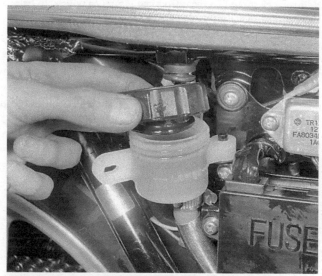

Rear brake master cylinder pivots out for cap removal

Verify transmission oil level and ...

... replenish if required

of each control cable. This will prevent drying-up of the cables between the more thorough lubrication that should be carried out during the 2000 mile/3 monthly service.

7 Rear chain lubrication and adjustment

In order that the life of the rear chain be extended as much as possible, regular lubrication and adjustment is essential.

Intermediate lubrication should take place at the weekly or 200 mile service interval with the chain in situ. Application of one of the proprietary chain greases contained in an aerosol can is ideal. Ordinary engine oil can be used, though owing to the speed with which it is flung off the rotating chain, its effectiveness is limited.

Adjust the chain after lubrication, so that there is approximately 20 mm (¾ in.) slack in the middle of the lower run. Always check with the chain at the tightest point as a chain rarely wears evenly during service.

Adjustment is accomplished after placing the machine on the centre stand and slackening the wheel nut, so that the wheel can be drawn backwards by means of the drawbolt adjusters in the fork ends. The torque arm nuts and the caliper bracket nut must also be slackened during this operation. Adjust the drawbolts an equal amount to preserve wheel alignment. The fork ends are clearly marked with a series of parallel lines above the adjusters, to provide a simple visual check.

8 Safety check

Give the machine a close visual inspection, checking for loose nuts and fittings, frayed control cables etc. Check the tyres for damage, especially splitting on the sidewalls. Remove any stones or other objects caught between the treads. This is particularly important on the front tyre, where rapid deflation due to penetration of the inner tube will almost certainly cause total loss of control!

9 Legal check

Ensure that the lights, horn and trafficators function correctly, also the speedometer.

---

3 monthly or every 2000 miles

Carry out the checks listed under the weekly/200 mile heading and then complete the following:

1 Cleaning and adjusting the contact breaker points

Remove the contact breaker inspection cover and gasket. The cover is retained by three screws. Inspect the faces of the two sets of contact breaker points. Slight pitting or burning can be removed while the contact breaker unit is in situ on the machine, using a very fine swiss file or emery paper (No. 400) backed by a thin strip of tin. If the pitting or burning is excessive, the contact breaker unit in question should be removed for points dressing or renewal (see Chapter 3, Section 4).

Rotate the engine until one set of points is in the fully open position. The correct gap is within the range 0.3 - 0.4 mm (0.012 - 0.016 in.). Adjustment is effected by slackening the screw holding the fixed contact breaker point in position and moving the point either closer or further away with a screwdriver inserted between the small upright post and the slot in the fixed contact plate. Make sure that the points are in the fully open position when this adjustment is made or a false reading will result. When the gap is correct, tighten the screw and re-check.

Repeat the procedure with the other set of points.

2 Checking and resetting the ignition timing

1 If the ignition timing is correct, the contact breaker points of the cylinder about to fire must be on the verge of separation when the piston is 2.3 mm (.090 in.) before top dead centre. An approximate indication of the accuracy of the timing is given by the small pointer in one of the apertures of the stator cover of

Aerosol lubricant is recommended for rear chain

Chain spring link must face direction of travel

Adjust chain by means of drawbolts

Check contact breaker points with feeler gauge

Adjust by slackening centre screw

Timing checked approximately by aligning scribe marks

the alternator. This should line up with a scribe mark on the face of the rotor when the contact breaker points of the cylinder involved are on the point of separation. It must be stressed that this is only an approximate indication of the accuracy of the setting. Optimum performance depends on timing the engine to a high degree of accuracy, to within $\pm$ 0.15 mm of the recommended setting on both cylinders.

2 To set the ignition timing with accuracy, remove the spark plug from the right-hand cylinder and fit a 14 mm dial gauge adaptor. Install the dial gauge and set it so that the dial shows a zero reading when the piston is exactly at top dead centre. Rotate the crankshaft backwards (clockwise) and check that the points for the right-hand cylinder (grey lead) are within the range 0.3 - 0.4 mm (0.012 in. - 0.016 in.) when they are fully open, then reverse the direction of rotation until the piston is 2.3 mm (0.090 in.) exactly before top dead centre. If the timing is correct, the contact breaker points should be about to separate.

3 If the timing is not correct, adjust the contact breaker points by moving them as a unit either clockwise or anti-clockwise, depending on whether the opening point needs to be advanced or retarded. This is accomplished by slackening the two cross head screws that pass through the elongated slots in the contact breaker base plate and moving the base plate with a screwdriver blade inserted between the short upright post and the notches in the edge of the base plate. When the adjustment is correct, tighten both screws and re-check the setting.

4 Repeat this procedure for the left-hand cylinder, adjusting the points to which the orange lead is attached. This setting must be made with equal accuracy.

5 As a final check, attach the positive lead of a 0-20 volt range dc voltmeter to each moving contact in turn and earth the negative lead. Switch on the ignition and check that the voltmeter commences to show a reading when the piston is 2.3 mm (0.090 in.) before top dead centre. Repeat for the other cylinder, with the other set of contact breaker points.

6 Before replacing the alternator cover, apply a few drops of light oil to the contact breaker cam wick. Do not over-lubricate or oil may find its way onto the points faces.

3 Cleaning the air filter

The air cleaner is located beneath the nose of the dual seat, which must be raised to gain access to the element. The lid of the air cleaner box is retained by two wing bolts which must be unscrewed before the lid is removed. The corrugated paper element can then be lifted out.

The element is cleaned by blowing through it from the inside with compressed air or by lightly tapping it so that loose dust on the surface will be displaced.

If the element is damaged or is contaminated badly in any way, it should be renewed. A blocked filter will increase fuel consumption and a perforated one may give rise to a weak mixture, resulting in a hot running engine. The filter is marked 'top' and 'front' and should be refitted accordingly.

Do not on any account run the machine with the air filter removed or with the air cleaner hoses disconnected. If this precaution is not observed, the engine will run with a permanently weak mixture, which will cause overheating and possible seizure.

4 Carburettor synchronisation

In order that the engine maintains the best possible performance at all times, the carburettors must always remain correctly adjusted and synchronised.

To check the synchronisation remove the adjusting orifice plugs from the right-hand side of each carburettor body. Operate the throttle twist grip until the throttle valves are fully open, and check that the lower edge of the aligning mark on each valve is aligned with the lower edge of the adjustment orifice on the respective carburettor (see photograph). If the adjustment is incorrect slacken the locknut on the throttle cable adjusters and adjust each carburettor independently. When adjustment is correct open and shut the throttle a number of

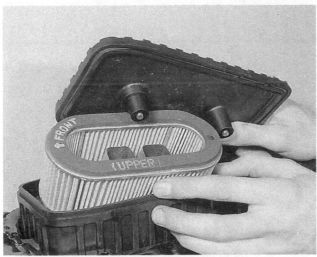

Air filter element is marked for correct installation

Unscrew the retaining screw to allow ...

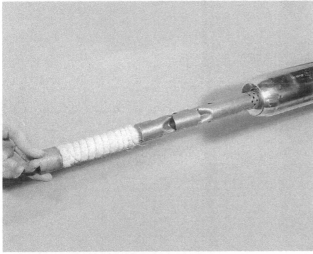

... removal of silencer baffle for cleaning

times and recheck the adjustment. Finally replace the orifice plugs and tighten the cable adjuster locknuts.

The idling speed and the mixture strength on each carburettor should now be adjusted as follows:

Remove the plug cap from the left-hand spark plug and start the engine. Screw in the throttle stop screw of the right-hand carburettor until the engine is running at a reasonably slow speed. Check whether adjustment of the pilot jet screw from the recommended setting of one and a half complete turns out, has any noticeable effect on even running and readjust the throttle stop screw as necessary. Note the tickover speed on this cylinder. Stop the engine.

Repeat the operation on the left-hand carburettor, with the right-hand spark plug removed. Adjust the idle speed so that it matches that of the right-hand cylinder. Run the engine with both plug caps reconnected. The tickover will probably be too fast, in which case the throttle stop screw on each carburettor should be screwed out exactly the same amount each and a little at a time, until the tickover speed is correct. Normally this is in the region of 1300 - 1400 rpm, as indicated by the tachometer.

5 Cleaning the exhaust system

Due to the oily nature of the exhaust gases of a two-stroke engine, the exhaust system will progressively block up with sludge and hard carbon deposits, as the machine is used. As the sludge builds up, back pressure will increase, with a resulting fall-off in performance. The deposits must therefore be removed at regular intervals.

To aid cleaning, the silencers are fitted with detachable baffles, which are retained by a single screw each, at the rear of each silencer. After removal of the screws, the baffles may be withdrawn by applying a stout pair of pliers to the crossbar in the baffle end, and twisting the baffle as necessary.

If the build-up of carbon and oil is not too great, a wash with a petrol/paraffin mixture will probably suffice as the cleaning medium. Otherwise, more drastic action will be necessary, such as the application of a blowlamp flame to burn away the accumulated deposits. After a number of cleaning operations, the fibreglass or asbestos wool with which the rearmost section of the baffle is shrouded, may disintegrate. Replacement of the wool is not strictly necessary as it will not affect performance but a slight increase in exhaust noise may be experienced.

When refitting the baffles, ensure that the retaining screws are tightened fully. If the screws fall out, the baffles will follow, creating excessive noise accompanied by a marked fall-off in performance. Running the engine without the baffles will also give rise to a weak mixture, causing overheating and possible engine seizure.

6 Oil pump adjustment

Synchronisation of the oil pump in relation to the carburettors should take place only **after** carburation synchronisation has been verified.

Commence by removing the alignment plug from the right-hand carburettor and detaching the oil pump cover, which is retained by three screws. Turn the throttle twist grip to full throttle, in which position the lower edge of the throttle valve mark should be aligned with the lower edge of the adjustment aperture. With the throttle in this position, the index mark on the outer face of the oil pump pulley should be exactly in line with the roll pin protruding from the pump plunger. If the alignment is not correct, loosen the locknut on the oil pump control cable adjuster and rotate the adjuster, as required, to bring the two marks into line. Tighten the locknut, open and shut the throttle a number of times and recheck the alignments.

7 Final drive chain lubrication

The final drive chain should be removed from the machine for thorough cleaning and lubrication if long service life is to be expected. This is in addition to the intermediate lubrication carried out with the chain on the machine, as described under the weekly/200 mile service heading.

With carburettor adjusted on full throttle as shown ...

... the oil pump pulley mark must align with boss pin

Apply a grease gun to the gearchange ...

... and the swinging arm pivot grease nipples

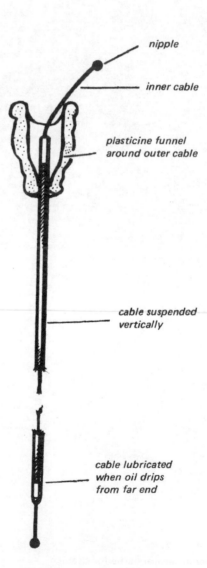

nipple

inner cable

plasticine funnel
around outer cable

cable suspended
vertically

cable lubricated
when oil drips
from far end

Fig. RM1. Control cable lubrication

Separate the chain by removing the master link and run it off the sprockets. If an old chain is available, interconnect the old and new chain, before the new chain is run off the sprockets. In this way the old chain can be pulled into place on the sprockets and then used to pull the regreased chain into place with ease.

Clean the chain thoroughly in a paraffin bath and then finally with petrol. The petrol will wash the paraffin out of the links and rollers which will then dry more quickly.

Allow the chain to dry and then immerse it in a molten lubricant such as Linklyfe or Chainguard. These lubricants must be used hot and will achieve better penetration of the links and rollers and are less likely to be thrown off by centrifugal force when the chain is in motion.

Refit the newly greased chain onto the sprocket, replacing the master link. This is accomplished most easily when the free ends of the chain are pushed into mesh on the rear wheel sprocket. The spring link must be fitted so that the closed end faces the normal direction of chain travel.

8 General lubrication

Apply a grease gun to the grease nipples on both ends of the swinging arm pivot shaft, until grease can be seen to exude from around the thrust bearing covers on either end of the cross-member. Wipe the excess grease off. Apply the gun to the gearchange shaft nipple, again wiping off any excess.

Apply grease or oil to the handlebar lever pivots and to the centrestand and propstand pivots.

9 Control cable lubrication

Lubricate the control cables thoroughly with motor oil or an all-purpose oil. A good method of lubricating the cables is shown in the accompanying illustration, using a plasticine funnel. This method has the disadvantage that the cables usually need removing from the machine. An hydraulic cable oiler which pressurises the lubricant overcomes this problem. Do not lubricate nylon lined cables (which may have been fitted as replacements), as the oil may cause the nylon to swell, thereby causing total cable seizure.

10 Changing the transmission oil

Place a container of greater than 1600 cc capacity below the gearbox and remove the oil filler plug from the right-hand engine cover. Unscrew the gearbox drain plug and allow the lubricant to drain. The oil should be drained with the engine hot, ideally after the machine has been on a run, as the lubricant will be thinner and so drain more rapidly and more completely.

When drainage is complete, refit the drain plug and the sealing washer and refill the gearbox with approximately 1500 cc of SAE 10W/30 engine oil. Allow the oil to settle and then check the level with the filler plug integral dipstick. When checking the level, do not screw the filler plug home, but allow it to rest on the orifice edge.

Six monthly or every 4000 miles

Complete the checks listed under the weekly/200 mile and three monthly/2000 mile headings, then complete the following additional procedures:

1 Changing the front fork damping oil

Place the machine on the centre stand so that the front wheel is clear of the ground. Place wooden blocks below the crankcase in order to prevent the machine from tipping forward. Loosen the pinch bolt, which clamps the top of each fork leg and remove the chrome cap bolts. Unscrew the drain plug from each fork leg, located directly above the wheel spindle, and allow the damping fluid to drain into a suitable container. This is accomplished most easily if the legs are attended to in turn. Take care not to spill any fluid onto the brake disc or tyre. The forks may be pumped up and down slowly to expel any remaining fluid. Refit and tighten the drain plugs. Refill each

fork leg with 145 cc of SAE 10W/30 engine oil, or a good quality fork oil. If a straight grade fork oil is chosen, either SAE 10, 20 or 30 may be used, depending on choice. The thicker the oil the heavier the damping will be. Refit and tighten the chrome cap bolts and then tighten the two pinch bolts.

2 Wheel condition (spoke type)

Check the spoke tension by gently tapping each one with a metal object. A loose spoke is identifiable by the low pitch noise generated. If any one spoke needs considerable tightening, it will be necessary to remove the tyre and inner tube in order to file down the protruding spoke end. This will prevent the spoke from chafing through the rim band and piercing the inner tube. Rotate the wheel and test for rim runout. Excessive runout will cause handling problems and should be corrected by tightening or loosening the relevant spokes. Care must be taken, since altering the tension in the wrong spokes may create more problems.

3 Cleaning the fuel tap

Because cleaning of the fuel tap requires that the tank be drained first, this operation should be carried out, if possible, when the fuel is at a low ebb, NOT at high tide! Disconnect the fuel feed pipe at the carburettor union after releasing the tension of the spring clip. Drain the fuel into a clean container, with the tap in the 'reserve' position.

Remove the filter (sediment trap) bowl from the main body of the fuel tap by applying a suitable spanner to the hexagon. Note the 'O' ring. The tap proper is retained on the underside of the tank by two screws passing through a flange. Support the tap as these are unscrewed. Clean the filter bowl and the filter matrix in clean fuel and then replace the components. If the condition of the 'O' ring or the tap/tank rubber seal is suspect, should be renewed.

4 Decarbonising the cylinder head, barrel and piston

Removal of the cylinder components, decarbonising and inspection, should be carried out by referring to the relevant Sections in Chapter 1. This work can be accomplished without removing the engine from the frame.

5 Removal, inspection and relubrication of wheel bearings

Carry out the operations listed in the heading by following the procedure given in Chapter 5, Section 9 for the front wheel and Section 11.2 for the rear wheel.

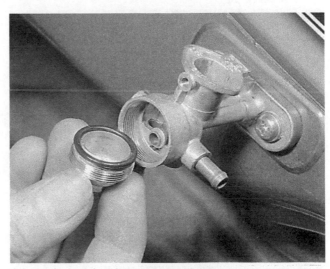

Unscrew fuel tap bowl to remove sediment

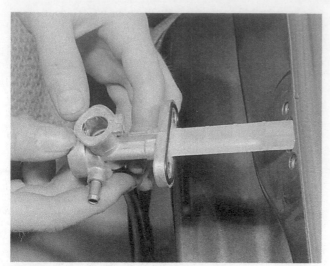

Withdraw the tap to clean filter column

Adjust clutch lifting mechanism clearance

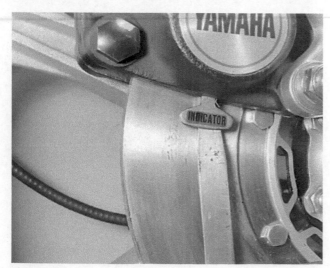

Check brake pad wear by means of feeler gauge

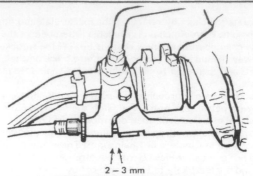

Fig. RM2. Clutch adjustment

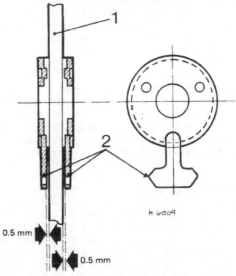

Fig. RM3. Checking brake pad wear with brake indicator tabs

1 Brake disc   2 Indicator

General maintenance adjustments

1 Clutch adjustment

The intervals at which the clutch should be adjusted will depend on the style of riding and the conditions under which the machine is used.

Adjust the clutch in two stages as follows:

Remove the clutch adjustment cover, which is retained by two screws. Loosen the cable adjuster screw locknut and turn the adjuster inwards fully, to give plenty of slack in the inner cable. Loosen the adjuster screw locknut in the casing and turn the screw clockwise until slight resistance is felt. Back off the screw about ¼ turn and tighten the locknut. The cover may be replaced.

Undo the cable adjuster screw at the handlebar lever, until there is approximately 2-3 mm (0.08 - 0.12 in.) play measured between the inner face of the lever and the stock face. (See accompanying diagram). Finally, tighten the cable adjuster locknut.

2 Checking brake pad wear

Brake pad wear depends largely on the conditions in which the machine is ridden and at what speed. It is difficult therefore, to give precise inspection intervals, but it follows that pad wear should be checked more frequently on a hard ridden machine.

The condition of each pad can be checked easily whilst still in situ on the machine. Each pad is fitted with an indicator tab; the distance between the tab and the brake disc indicating the severity of wear. Measure the gap with a feeler gauge. If the gap on any indicator is less than 0.5 mm (0.1968 in.) both pads in that set should be renewed. It is not necessary to replace the front and the rear brake pads at the same time, unless both sets are worn. In practice the front pads will wear at approximately three times the rate of the rear pads.

Worn pads may be eased from position in the caliper, after removal of the relevant wheel. When refitting new pads, ensure that the anti-chatter plate is fitted correctly to the rear of each pad. Do not apply the brake when the pads are out or when the disc is not between the pads. If this precaution is not observed, the piston may be forced out, causing fluid loss, and in any case difficulty will be experienced in replacing the wheel.

Do not expect the brake to work perfectly after pad renewal. Care should be taken for the first 50 miles, to allow the pads to bed down on the disc.

Quick glance:
Maintenance adjustments and capacities

ENGINE	SAE 30 two-stroke oil. Separate lubrication system operated by 'Autolube' oil pump, interconnected with the throttle
OIL TANK	Capacity 1.9 US qt (1.8 litre)
GEARBOX	Capacity 2.6 Imp. pints (1.5 litre). SAE 10W/30 engine oil
FRONT FORKS	145 cc per fork leg SAE 10W/30 engine oil
CONTACT BREAKER GAP	0.3 to 0.4 mm (0.012 in to 0.016 in)
SPARK PLUG GAP	0.6 - 0.7 mm (0.024 - 0.028 in)
TYRE PRESSURES:	
Solo	26 psi front, 28 psi rear
Pillion, or high speed	28 psi front, 33 psi rear

Working conditions and tools

When a major overhaul is contemplated, it is important that a clean, well-lit working space is available, equipped with a workbench and vice, and with space for laying out or storing the dismantled assemblies in an orderly manner where they are unlikely to be disturbed. The use of a good workshop will give the satisfaction of work done in comfort and without haste, where there is little chance of the machine being dismantled and reassembled in anything other than clean surroundings. Unfortunately, these ideal working conditions are not always practicable and under these latter circumstances when improvisation is called for, extra care and time will be needed.

The other essential requirement is a comprehensive set of good quality tools. Quality is of prime importance since cheap tools will prove expensive in the long run if they slip or break and damage the components to which they are applied. A good quality tool will last a long time, and more than justify the cost. The basis of any tool kit is a set of open-ended spanners, which can be used on almost any part of the machine to which there is reasonable access. A set of ring spanners makes a useful addition, since they can be used on nuts that are very tight or where access is restricted. Where the cost has to be kept within reasonable bounds, a compromise can be effected with a set of combination spanners - open-ended at one end and having a ring of the same size on the other end. Socket spanners may also be considered a good investment, a basic 3/8 in. or 1/2 in. drive kit comprising a ratchet handle and a small number of socket heads, if money is limited. Additional sockets can be purchased, as and when they are required. Provided they are slim· in profile, sockets will reach nuts or bolts that are deeply recessed. When purchasing spanners of any kind, make sure the correct size standard is purchased. Almost all machines manufactured outside the UK and the USA have metric nuts and bolts, whilst those produced in Britain have BSF or BSW sizes. The standard used in the USA is AF, which is also found on some of the later British machines. Other tools that should be included in the kit are a range of crosshead screwdrivers, a pair of pliers and a hammer.

When considering the purchase of tools, it should be remembered that by carrying out the work oneself, a large proportion of the normal repair cost, made up by labour charges, will be saved. The economy made on even a minor overhaul will go a long way towards the improvement of a tool kit.

In addition to the basic tool kit, certain additional tools can prove invaluable when they are close to hand, to help speed up a multitude of repetitive jobs. For example, an impact screwdriver will ease the removal of screws that have been tightened by a similar tool, during assembly, without risk of damaging the screw heads. And, of course, it can be used again to retighten the screws, to ensure an oil or airtight seal results. Circlip pliers have their uses too, since gear pinions, shafts and similar components are frequently retained by circlips that are not too easily displaced by a screwdriver. There are two types of circlip plier, one for internal and one for external circlips. They may also have straight or right-angled jaws.

One of the most useful of all tools is the torque wrench, a form of spanner that can be adjusted to slip when a measured amount of force is applied to any bolt or nut. Torque wrench settings are given in almost every modern workshop or service manual, where the extent to which a complex component, such as a cylinder head, can be tightened without fear of distortion or leakage. The tightening of bearing caps is yet another example. Overtightening will stretch or even break bolts, necessitating extra work to extract the broken portions.

As may be expected the more sophisticated the machine, the greater is the number of tools likely to be required if it is to be kept in first class condition by the home mechanic. Unfortunately, there are certain jobs which cannot be accomplished successfully without the correct equipment and although there is invariably a specialist who will undertake the work for a fee, the home mechanic will have to dig more deeply in his pocket for the purchase of similar equipment if he does not wish to employ the services of others. Here a word of caution is necessary, since some of these jobs are best left to the expert. Although an electrical multimeter of the AVO type will prove helpful in tracing electrical faults, in inexperienced hands it may irrevocably damage some of the electrical components if a test current is passed through them in the wrong direction. This can apply to the synchronisation of twin or multiple carburettors too, where a certain amount of expertise is needed when setting them up with vacuum gauges. These are, however, exceptions. Some instruments, such as a strobe lamp, are virtually essential when checking the timing of a machine powered by CDI ignition system. In short, do not purchase any of these special items unless you have the experience to use them correctly.

Although this manual shows how components can be removed and replaced without the use of special service tools (unless absolutely essential), it is worthwhile giving consideration to the purchase of the more commonly used tools if the machine is regarded as a long term purchase. Whilst the alternative methods suggested will remove and replace parts without risk of damage, the use of the special tools recommended and sold by the manufacturer will invariably save time.

Chapter 1 Engine clutch and gearbox

Contents

Specifications

Engine

Type	Parallel twin cylinder two-stroke
Porting	6 port
Capacity	398 cc
Bore	64 mm (2.51 in)
Stroke	62 mm (2.44 in)
Compression ratio	6.2 : 1
bhp	40 (DIN) @ 7000 rpm
Lubrication	Separate oil pump using 'Autolube' system

Cylinder barrel

Bore size (standard)	64 + 0.02 mm (2.51 + 0.0008 in)
Taper limit	0.05 mm (0.002 in)
Ovality limit	0.01 mm (0.0004 in)

Pistons

Clearance in bore	0.035 - 0.040 mm (0.0013 - 0.0015 in)
Oversizes available	+ 0.25, + 0.50, + 0.75, + 1.0 mm (0.0095, 0.019. 0.028. 0.039 in)

Piston rings

No. per piston	...	...	...	...	...	...	...	...	2
Type:									
Top	...	...	...	...	...	...	...	...	Dykes
Second	...	...	...	...	...	...	...	...	Plain
End gap	...	...	...	...	...	...	...	...	0.3 - 0.5 mm (0.012 - 0.019 in)

Gearbox

Type	...	...	...	...	...	...	...	...	6-speed constant mesh
Ratios:									
1st gear	...	...	...	...	...	...	...	...	2.571 : 1
2nd gear	...	...	...	...	...	...	...	...	1.778 : 1
3rd gear	...	...	...	...	...	...	...	...	1.318 : 1
4th gear	...	...	...	...	...	...	...	...	1.083 : 1
5th gear	...	...	...	...	...	...	...	...	0.962 : 1
6th gear	...	...	...	...	...	...	...	...	0.889 : 1

Clutch

Type	...	...	...	...	...	...	...	...	Wet, multi-plate
No. of plates:									
Friction	...	...	...	...	...	...	...	...	7
Plain	...	...	...	...	...	...	...	...	6
Clutch springs:									
No. of	...	...	...	...	...	...	...	...	6
Free length	...	...	...	...	...	...	...	...	36.4 mm (1.423 in)
Wear limit	...	...	...	...	...	...	...	...	35.4 mm (1.393 in)
Friction plate thickness	...	...	...	...	...	...	...	...	3.0 mm (0.1181 in)
Wear limit	...	...	...	...	...	...	...	...	2.7 mm (0.1062 in)

1 General description

The engine unit fitted to the Yamaha RD 400 model is a twin cylinder two-stroke, using flat top pistons and what is known as the loop scavenging system to effect a satisfactory induction and exhaust sequence. Each cylinder barrel has six ports (inlet, exhaust, two transfer and two auxiliary transfer) the design of which permits the incoming charge to expel the exhaust gases and at the same time cool the piston. The piston rings are pegged in characteristic two-stroke practice, to prevent them from rotating and the ends being trapped in the cylinder barrel ports. Large diameter oil seals form an effective seal around the built-up crankshaft assembly, which runs on four journal ball bearings and has full flywheels. All engine/gear castings are in aluminium alloy, including the cylinder barrels and cylinder heads. The crankcase is arranged to separate horizontally, simplifying dismantling and reassembly operations. Because the engine and gearbox are built in unit, when the engine is dismantled the gearbox has to be dismantled too, and vice-versa.

The generator is on the left-hand side of the engine. It is an alternator with the rotor attached to the end of the crankshaft. The clutch assembly is located on the right-hand side, behind the cover containing the kickstarter mechanism. The exhaust system is twin downswept, and each cylinder has its own exhaust pipe and silencer. The gearchange pedal is on the left-hand side of the machine; it operates a five-speed constant mesh gearbox.

Lubrication is effected by the Yamaha "Autolube" system, which takes the form of a gear driven oil pump drawing oil from a separate oil tank and distributing it to the various working parts of the engine. The pump is interconnected to the throttle, so that optimum lubrication is achieved at all times, corresponding to the requirements of both engine speed and throttle opening. This system completely obviates the need for petroil and the problems that arise when pre-mixing petrol and oil.

2 Operations with engine in frame

It is not necessary to remove the engine unit from the frame unless the crankshaft assembly and/or the gearbox internals require attention. Most operations can be accomplished with the engine in place, such as:

1 Removal and replacement of the cylinder heads.
2 Removal and replacement of the cylinder barrels and pistons.
3 Removal and replacement of the generator.
4 Removal and replacement of the clutch.
5 Removal and replacement of the contact breaker assembly.

When several operations need to be undertaken simultaneously, it will probably be advantageous to remove the complete engine unit from the frame, an operation that should take approximately two hours, working at a leisurely pace. This will give the advantage of better access and more working space.

3 Operations with engine removed

1 Removal and replacement of the crankshaft assembly.
2 Removal and replacement of the gear cluster, selectors and gearbox main bearings.

4 Method of engine/gearbox removal

As mentioned previously, the engine and gearbox are built in unit and it is necessary to remove the unit complete, in order to gain access to components. Separation is accomplished after the engine unit has been removed from the frame and refitting cannot take place until the crankcase has been reassembled. When the crankcase is separated the gearbox internals will be exposed.

5 Removing the engine/gear unit

1 Place the machine on the centre stand and make sure that it is standing firmly on level ground.
2 Turn the fuel tap to the position marked 'stop'. Disconnect the two fuel pipes where they join the float chamber of each carburettor. The pipes are a push-on fit. If the tank still contains fuel, drain it off by turning the tap to the 'on' position. When the tank is drained, disconnect the rubber pipe connecting both halves of the petrol tank, on the underside. The tank can now be lifted away from the machine; it is retained in position by rubber buffers at the front and by two bolts passing through projections to the rear of the tank. Remove the bolts, noting the buffers

5.2 Petrol tank secured by two bolts at rear

5.4a Detach the exhaust pipe flanges and ...

5.4b ... pull each pipe from the connector

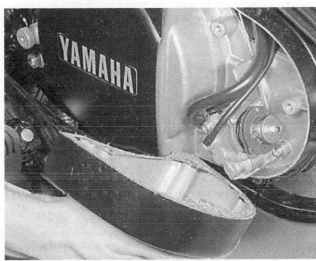

5.5 Oil pump cover held by three screws

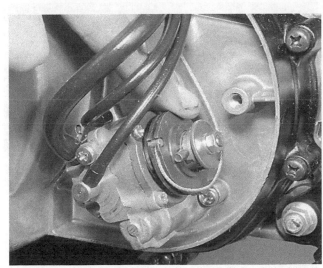

5.6 Detach oil pump control cable

through which they pass. Lift the tank up at the rear and ease it backwards carefully, until it comes clear.

3   Although there is no necessity to detach the petrol tank in order to remove the engine/gear unit, better access will be gained, with less risk of damage to the painted surfaces, if the tank is out of the way.

4   Remove the nuts from each exhaust pipe flange at the joint between the exhaust pipe and cylinder barrel. Each exhaust pipe is a push fit in the rubber connector piece. Pull the pipes and connectors from position. The silencers may be left in place on the machine.

5   Remove the three cross-head screws retaining the front portion of the right-hand crankcase cover; detach the cover. This will give access to the oil pump. Release the wire clip around the oil feed pipe and pull the pipe from the oil pump connection. Allow the oil tank to drain, or alternatively block the end of the pipe with a bolt of convenient size to impede the flow of oil, whilst dismantling continues.

6   Disconnect the control cable linking the oil pump with the throttle. A barrel-shaped nipple on the end of the cable engages with the oil pump pulley and can be slipped out of position by twisting, whilst the pulley is turned. When the cable is free, remove the adjuster which threads into the outer crankcase cover. The cable can then be withdrawn completely from the engine.

7   The electrical connections are found within a plastic sleeve close to the air intake of the left-hand carburettor. Separate the connectors; note the colour coding.

8   Remove the gear change lever on the left-hand side of the machine. It is retained on a splined shaft by a pinch bolt; when the pinch bolt is removed the lever can be pulled off the splines. Mark both the lever and the splined shaft. The lever can then be replaced in the same position.

9   Remove the three cross-head screws securing the generator cover on the left-hand side of the machine; remove the cover, noting that it is located on dowels and is therefore a tight fit. The outer crankcase cover, also secured by three cross-head screws, can now be detached. Before the crankcase cover can be lifted clear it will be necessary to detach the clutch cable, from the clutch operating arm within the cover. Note how the lower end of the cable fits into an abutment attached to a small portion of the crankcase, retained in position by the outer cover.

10 Withdraw the contact breaker assembly complete with the brush gear cover enclosing the generator. This is retained by three cross-head screws around the periphery. The assembly can be withdrawn, complete with wiring harness, if the rubber grommet is detached from the back of the upper crankcase casting. The circular aperture will provide sufficient clearance for the cable connectors to pass through. Note that the separate lead to the neutral contact must be detached by removing the single screw which passes through the terminal at the neutral indicator switch.

11 Detach the contact breaker cam from the generator armature. When the centre retaining bolt is unscrewed, the cam is freed and can be lifted away. It is located by means of a peg and slot arrangement, so that it will always be replaced in the correct position.

12 A Yamaha service tool is recommended to remove the generator armature from the end of the crankshaft. This takes the form of a puller bolt that will release the armature from the tapered end of the crankshaft, when it is screwed home. If the service tool is not available, the same effect can be achieved by using the rear uppermost engine mounting bolts in its place. The taper joint is keyed; remove the key and place it in a safe place for subsequent reassembly.

13 Part the final drive chain at the spring connecting link and remove the chain for later inspection and lubrication.

14 Detach the air cleaner hoses from both carburettors. The hoses are retained around each carburettor bell mouth by a screw clip. Remove both hoses; the other end is a push fit over the outlets from the air cleaner assembly, and is also retained by a clip.

15 Disconnect the tachometer drive cable by unscrewing the gland nut. This is found to the rear of both carburettors, close to

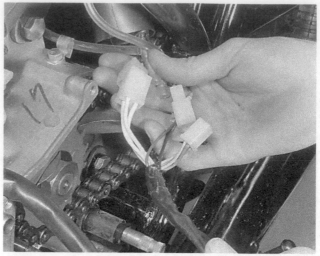

5.7a Separate the main electrical connectors and ...

5.7b ... disengage the cable from the clip

5.9 Clutch cable must be detached from cover

5.10a Remove complete stator held by three screws

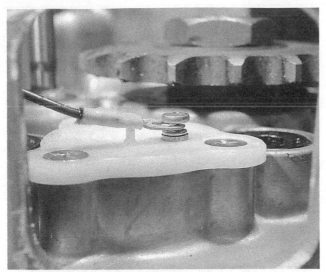

5.10b Detach neutral warning switch lead

5.10c Alternator wires may be pulled through casing hole

5.11 Unscrew centre bolt and remove contact breaker cam

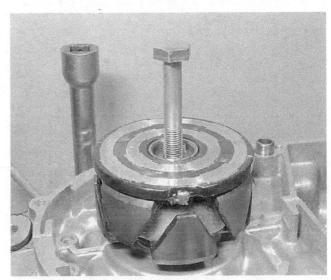

5.12 Extract alternator rotor using engine mounting bolt

5.14 Air cleaner hose is retained by screw clips

5.15 Unscrew knurled ring to detach tachometer cable

the top engine mounting bolt.

16 Unscrew both carburettor tops and remove them, complete with their needle and throttle assembly. These parts are very easily damaged and should be taped out of harms way to some other part of the machine.

17 Detach the oil feed pipe which runs from the oil pump to the carburettors. Each pipe is secured by a spring collar at the union. The carburettors can be pulled away from their flexible mountings if each clip fitting is unscrewed first. Slacken the cross-head screw that passes through each clip.

18 Remove both spark plug leads by pulling off the plug caps.

19 Remove the four engine mounting bolts and the rear mounting plates attached to the lug on the upper crankcase, close to the tachometer drive take-off. The engine unit can now be lifted from the frame, preferably from the right-hand side. Although the engine unit is not unduly heavy, it is advisable to have a second person available during this operation, if only to steady the machine and ensure there is sufficient clearance as the engine is lifted from position.

6 Dismantling the engine and gearbox: general

1   Before commencing work on the engine unit, the external

5.17 Disconnect oil feed pipes to carburettors

5.19a Detachable engine plate at upper rear and ...

5.19b ... lower right-hand front mounting points

5.19c Lower rear bolt passes through fixed lugs

5.19d Lift engine out from right-hand side

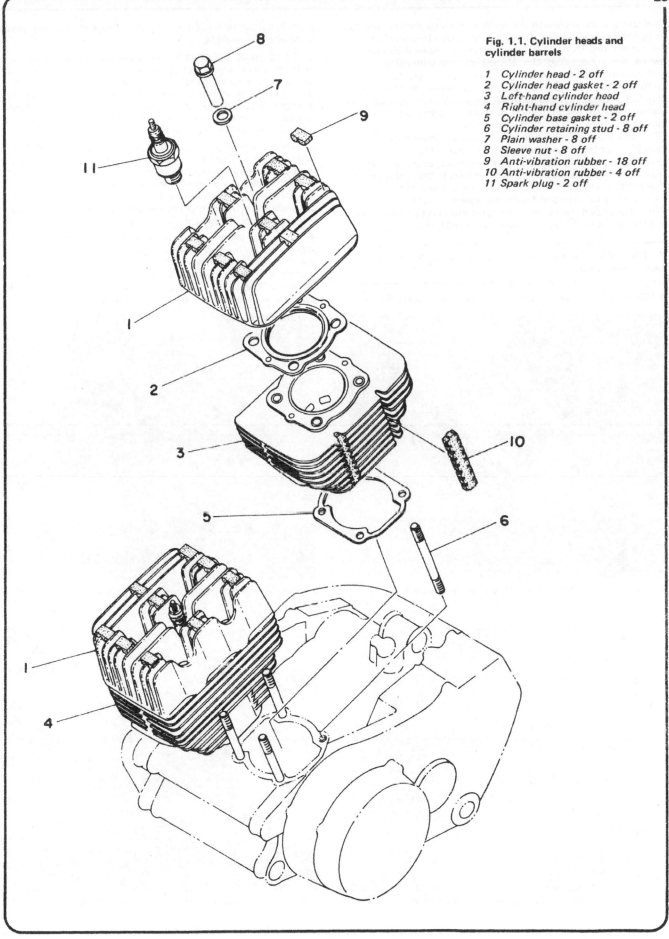

Fig. 1.1. Cylinder heads and cylinder barrels

1 Cylinder head - 2 off
2 Cylinder head gasket - 2 off
3 Left-hand cylinder head
4 Right-hand cylinder head
5 Cylinder base gasket - 2 off
6 Cylinder retaining stud - 8 off
7 Plain washer - 8 off
8 Sleeve nut - 8 off
9 Anti-vibration rubber - 18 off
10 Anti-vibration rubber - 4 off
11 Spark plug - 2 off

surfaces should be cleaned thoroughly. A motor cycle engine has very little protection from road grit and other foreign matter, which will find its way into the dismantled engine if this simple precaution is not taken. One of the proprietary cleaning compounds, such as "Gunk" or "Jizer" can be used to good effect, particularly if the compound is permitted to work into the film of oil and grease before it is washed away. Special care is necessary when washing down, to prevent water from entering the now exposed parts of the engine unit.

2   Never use undue force to remove any stubborn part unless specific mention is made of this requirement. There is invariably good reason why a part is difficult to remove, often because the dismantling operation has been tackled in the wrong sequence. Dismantling will be made easier if a simple engine stand is constructed to correspond with the engine mounting points. This arrangement will permit the complete unit to be clamped rigidly to the workbench, leaving both hands free.

7   Dismantling the engine unit: removing the cylinder heads, barrels and pistons

1   Each cylinder head is retained by four sleeve bolts. Slacken

them in a diagonal sequence and then remove the cylinder heads, complete with gaskets.

2   Withdraw the cylinders from the crankcase, one at a time, by lifting them upwards along the holding down studs. Take care to support each piston and rings as it emerges from the cylinder bore; it is a wise precaution to place some clean rag in the mouth of each crankcase before the piston is released from the cylinder so that in the event of piston ring breakages, particles of broken ring will not fall into the crankcase.

3   Remove the circlips from each piston and press out the gudgeon pins so that the pistons are released. If the gudgeon pins are a particularly tight fit, the pistons should be warmed first, to expand the alloy and release the grip on the steel pins. If it is necessary to tap a gudgeon pin out of position, make sure the connecting rod is supported to prevent distortion. On no account use excess force. Discard the old circlips.

4   Before the pistons are placed aside for examination at a later stage, mark each inside the skirt so that it is replaced in an identical position. It is advisable to keep the gudgeon pins 'matched' with the pistons from which they were originally extracted. There is no need to mark the back and front of each piston as the front is denoted by an arrow on the piston crown.

7.1 Cylinder head/barrel retained by four sleeve nuts

7.2 Carefully lift the barrel upwards off the studs

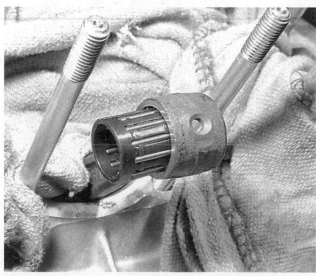

7.3a Remove circlip and then push out gudgeon pin

7.3b Small-end bearing is a push fit in connecting rod

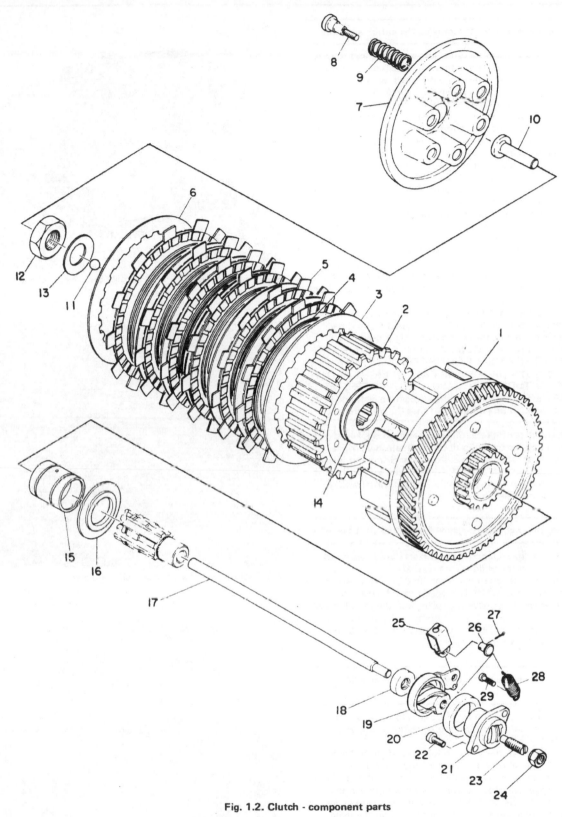

Fig. 1.2. Clutch - component parts

1	Primary driven gear and clutch outer drum	7	Pressure plate	15	Spacer
2	Clutch inner drum	8	Clutch spring screw - 6 off	16	Thrust washer
3	Plain clutch plate	9	Clutch spring - 6 off	17	Pushrod
4	Cushion ring - 7 off	10	Clutch pushrod 'mushroom'	18	Pushrod seal
5	Friction (inserted) clutch plate - 7 off	11	Ball bearing (5/16 in. dia.)	19	Clutch actuating mechanism
6	Plain clutch plate - 6 off	12	Locknut	20	Dust seal
		13	Belville spring washer	21	Actuating mechanism housing
		14	Thrust washer	22	Screw - 2 off

23	Adjusting screw
24	Adjusting nut
25	Trunnion for cable
26	Clevis pin
27	Split pin
28	Return spring
29	Spring anchor

8 Dismantling the engine unit: removing the clutch assembly

1   Lay the crankcase assembly on its left-hand side and remove the right-hand crankcase cover. Before the cover can be removed, it will first be necessary to detach the kickstart, which is retained on a splined shaft by a pinch bolt. Slacken the pinch bolt and draw the kickstart off the splines.

2   The crankcase cover is held in position by nine cross-head screws and two dowels. When the screws are removed, the cover will pull away, complete with the oil pump, after the oil pump cable adjuster is unscrewed from the top. There is no necessity to detach the oil pump as a separate unit, unless renewal is necessary.

3   The clutch assembly is now exposed. Commence by unscrewing the six hexagon headed screws in the pressure plate and remove them, together with the six clutch springs. If the pressure plate is lifted away, the clutch plates can be withdrawn. There are thirteen of these, six plain plates and seven with friction inserts, alternatively spaced. Note the presence of rubber cushion rings, one of which is placed between each friction plate and plain plate. There are seven rings altogether. Remove the mushroom-headed portion of the clutch rod assembly which protrudes from the hollow mainshaft.

4   The clutch centre is retained by a very tight nut. The use of Yamaha service tool 90890 - 01024 is advised to hold the clutch centre firmly whilst the centre nut is slackened by a socket spanner. ON NO ACCOUNT ATTEMPT TO LOCK THE CLUTCH CENTRE BY MEANS OF THE OUTER DRUM SERRATIONS; BOTH DRUMS ARE CAST IN LIGHT ALLOY AND WILL FRACTURE VERY EASILY IF OVERSTRESSED IN THIS MANNER. Alternatively, engage bottom gear and hold the final drive sprocket retaining nut so that the mainshaft is locked in position.

5   When the clutch centre retaining nut has been slackened and removed, the clutch centre will pull off the mainshaft splines, followed by a thrust washer. The clutch outer drum and integral primary drive gear will also pull off. Remove the spacer that fits within the clutch outer drum assembly and the thrust washer that precedes the mainshaft oil seal.

9 Dismantling the engine unit: removing the kickstart assembly

1   Remove the circlip retaining the kickstart double gear idler pinion, and withdraw the pinion from its shaft.

2   The kickstart ratchet assembly is retained by the kickstart return spring, one end of which is looped around a projection cast into the crankcase. Remove the spring and then the kickstart assembly as a complete unit. There is no necessity to dismantle the unit further, unless replacements are required. See Section 24 of this Chapter for further details.

10 Dismantling the engine unit: removing the primary drive pinion

1   Lock the engine in position by placing a stout metal rod through the eye of both connecting rods, resting it on two wooden supports across the crankcase.

2   Unscrew and remove the pinion retaining nut (right-hand thread) and pull the pinion off the end of the crankshaft. It is located by means of a key, which should be removed and placed in a safe place, together with the domed washer between the retaining nut and the pinion. No difficulty should be experienced in removing the pinion because it is a parallel fit on the mainshaft.

11 Dismantling the engine unit: removing the gear change mechanism

1   Replace the engine in the upright position, resting on the

8.2 Primary drive cover held by nine cross-head screws

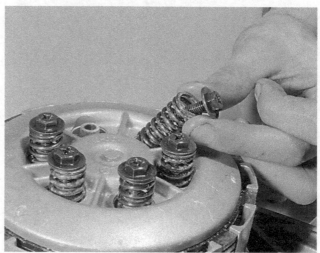

8.3 Remove clutch springs and all the plates

8.4 The clutch centre is retained by a nut and tab washer

8.5 Lift the outer drum from position

9.1 Kickstart idler pinion held by a circlip

9.2a Disengage the kickstart spring and ...

9.2b ... remove the kickstart assembly as a unit

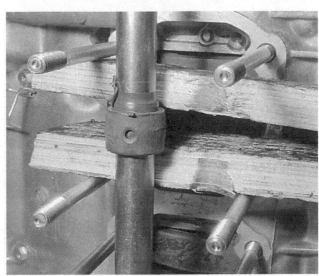

10.1 Lock the crankshaft as shown and ...

10.2 ... remove the primary drive pinion and key

bottom of the crankcase. Remove the rubber cover over the gear change shaft on the left-hand side and the circlip and shim washer that lie beneath. The complete shaft can now be pulled out from the right-hand side of the engine unit. A slot in the arm attached to the end of the shaft engages with a peg on the operating arm of the gear change mechanism. Do not lose the shouldered collar that slides over the peg.

2   Remove the circlip from the gear selector operating arm and lift the arm away from the spindle, on which it pivots. It will be necessary to open out the two spring-loaded pawls so that they will disengage from the pins of the gear selector drum. The operating arm will come away complete with the return spring.

12 Dismantling the engine unit: removing the final drive sprocket

1   Working from the left-hand side of the engine unit, hold the final drive sprocket with the appropriate Yamaha service tool or alternatively wrap the final drive chain around the sprocket and secure both ends in a vice so that the sprocket is prevented from moving. Alternatively, the chain may be 'bunched' against the casing to effect locking. Care must be taken if this method is used, to prevent damage to the case.

2   Bend back the tab washer that locks the sprocket retaining nut and unscrew the nut. It is best to use a good fitting socket spanner for this purpose since the nut is very tight and may require considerable leverage to move it initially. When the nut is removed, the sprocket will pull off the shaft without difficulty.

13 Dismantling the engine unit: separating the crankcase

1   Invert the crankcase and remove the eight nuts and three bolts each of which is numbered. The nuts and bolts should be slackened in numerical order, commencing with the highest number and working downward.

2   When all the bottom nuts, bolts and washers have been detached, turn the crankcase upright again and remove the eight bolts on the top surface. These too are numbered and should be slackened and removed in similar order.

3   When all nineteen nuts and bolts have been withdrawn, the crankcase can be separated. Part them by striking the front part of the upper crankcase and the rear part of the lower crankcase with a soft-faced mallet. Tap gently - little force is needed to effect the separation.

14 Dismantling the engine unit: removing the crankshaft assembly and gear clusters

1   Remove the crankshaft by tapping each end with a soft-faced mallet. It will come away as a complete unit, together with the three oil seals, one at each end of the crankshaft and one in the centre, between the two centre ball races.

2   The gear clusters will lift out in similar fashion, together with their respective shafts, bearings and oil seals.

3   The selector rods on which the selector forks pivot are a sliding fit in the lower crankcase and can be pushed out from the right-hand side. Both have an oil seal on the left-hand side, which

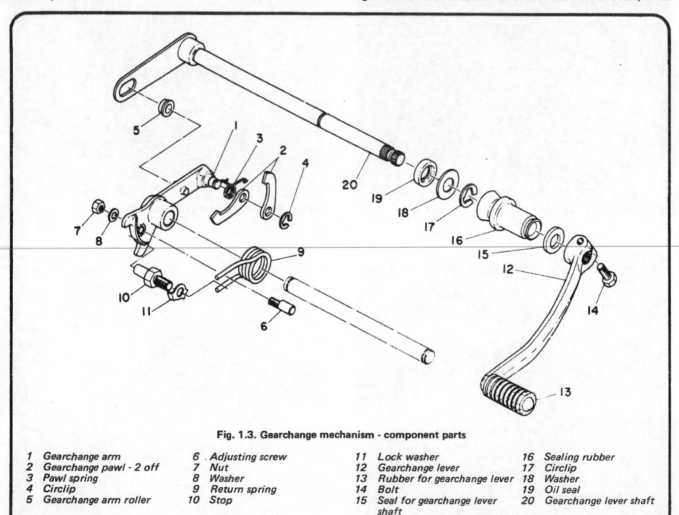

Fig. 1.3. Gearchange mechanism - component parts

1 Gearchange arm	6 Adjusting screw	11 Lock washer	16 Sealing rubber
2 Gearchange pawl - 2 off	7 Nut	12 Gearchange lever	17 Circlip
3 Pawl spring	8 Washer	13 Rubber for gearchange lever	18 Washer
4 Circlip	9 Return spring	14 Bolt	19 Oil seal
5 Gearchange arm roller	10 Stop	15 Seal for gearchange lever shaft	20 Gearchange lever shaft

11.1a Prise the 'E' clip from the gearchange shaft and ...

11.1b ... withdraw the shaft from the right-hand side

11.2 Remove 'E' clip to free change pawl arm

12.1 Knock down the 'ear' of the tab washer

12.2 Lock sprocket to enable loosening of nut

14.1 Support connecting rods when lifting crankshaft

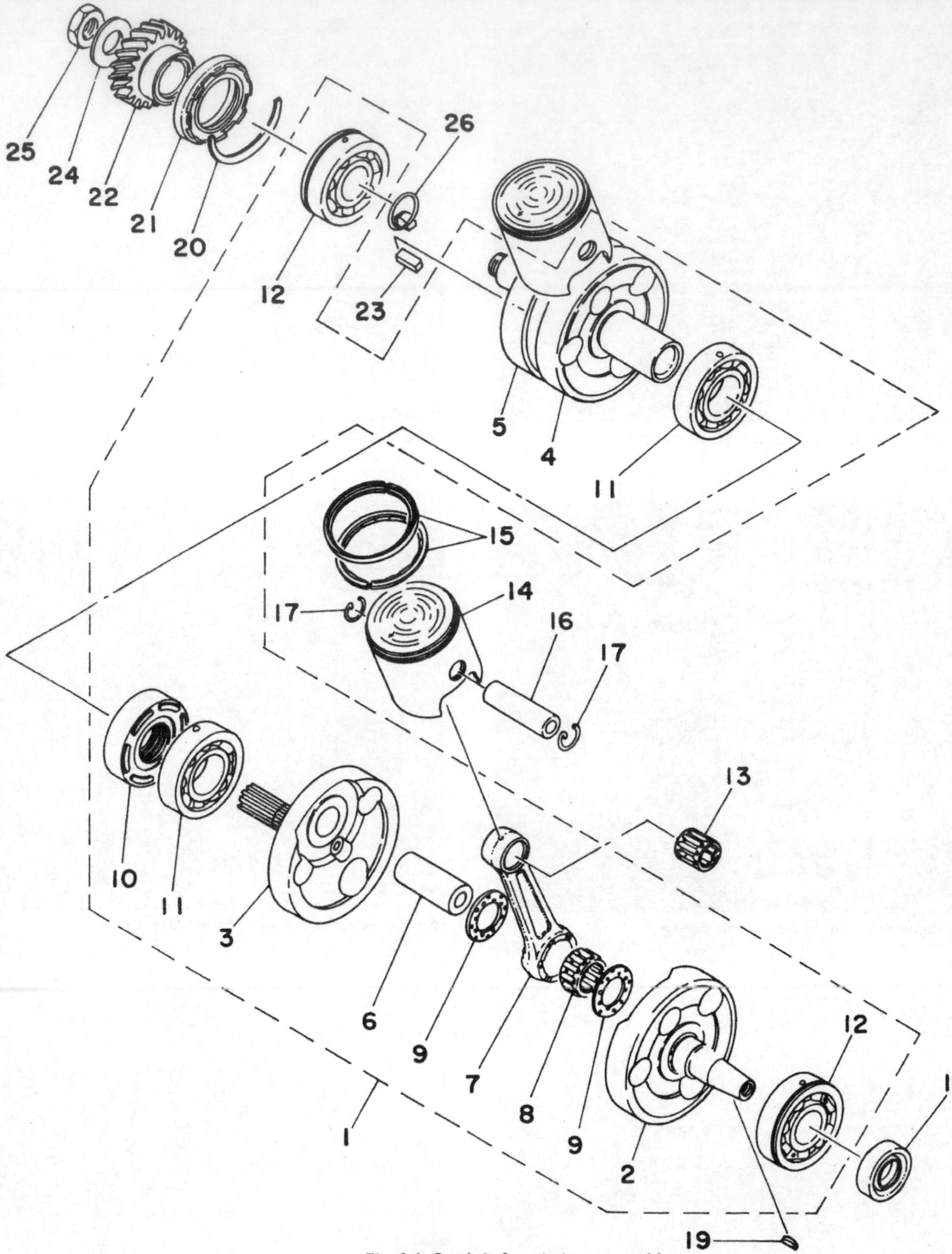

Fig. 1.4. Crankshaft and piston assembly

1 Crankshaft assembly	10 Labyrinth seal	19 Woodruff key
2 LH outer flywheel	11 Main bearing - 2 off	20 Bearing half clip
3 LH inner flywheel	12 Main bearing - 2 off	21 Oil seal
4 RH inner flywheel	13 Small end bearing - 2 off	22 Primary drive pinion (23T)
5 RH outer flywheel	14 Piston - 2 off	23 Straight-cut key
6 Crankpin - 2 off	15 Piston ring set - 2 off	24 Belville washer
7 Connecting rod - 2 off	16 Gudgeon pin - 2 off	25 Nut
8 Big-end bearing - 2 off	17 Gudgeon pin circlip - 4 off	26 Special 'O' ring
9 Thrust washer - 4 off	18 Oil seal	

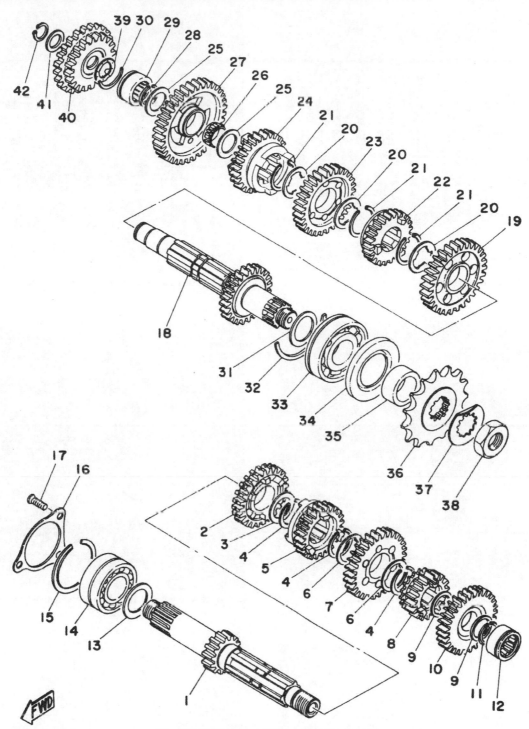

Fig. 1.5. Gearbox assembly - component parts

1 Mainshaft (14T)
2 Mainshaft 4th gear pinion (24T)
3 Thrust washer
4 Circlip - 3 off
5 Mainshaft 3rd gear pinion (22T)
6 Thrust washer - 2 off
7 Mainshaft 6th gear pinion (27T)
8 Mainshaft 2nd gear pinion (18T)
9 Thrust washer - 2 off
10 Mainshaft 5th gear pinion (27T)
11 Circlip
12 Needle roller bearing
13 Shim
14 Journal ball bearing
15 Bearing clip
16 Bearing retainer plate
17 Countersunk screw - 3 off
18 Layshaft (25T)
19 Layshaft 2nd gear pinion (32T)
20 Splined washer - 3 off
21 Circlip - 3 off
22 Layshaft 6th gear pinion (24T)
23 Layshaft 3rd gear pinion (29T)
24 Layshaft 4th gear pinion (26T)
25 Thrust washer - 2 off
26 Needle roller bearing
27 Layshaft 1st gear pinion (36T)
28 Circlip
29 Needle roller bearing
30 Bearing half-clip
31 Mainshaft shim - A/R
32 Circlip
33 Journal ball bearing
34 Oil seal
35 Dostance collar
36 Final drive sprocket (16T or17T) - A/R
37 Tab washer
38 Centre nut
39 Wave washer
40 Kickstart/oil pump idler gear (22T and 27T)
41 Shim
42 Circlip

14.2a Gear clusters - general view

14.2b Lift out each shaft assembly separately

14.3 Remove circlips and drift selector rods out

15.1 Access to neutral switch terminal gained by removing cover

15.3a Tachometer pinion retained by an 'E' clip

15.3b Tachometer shaft housing is secured by a claw

must be removed before the selector rods can be freed. A circlip at the left-hand end of each rod prevents their displacement from the left. The selector rod carrying the two rearmost selector forks has a circlip inside the crankcase at the left-hand end. This must be removed. The front rod has a circlip on the right-hand end which must also be removed.

4   When the selector rods have been removed, the selector forks will be free to lift away since the rods pass through them. Mark the forks so that they are replaced in the same positions; both pairs of selector forks are 'handed'.

5   Remove the circlip halves used to locate the various ball races in the engine and gearbox assembly, and place them aside until reassembly commences.

15 Dismantling the engine unit: removing the neutral switch, tachometer drive pinion and oil pump drive pinion

1   There is no necessity to remove the neutral indicator switch unless the switch malfunctions or the gear selector drum is to be removed. The switch unit is retained to the crankcase by three cross-head screws which, when withdrawn, will release the switch unit.

2   If the gear selector drum is to be removed, unscrew the cross-head screw in the centre of the end of the drum and detach the circular end plate together with the neutral contact and spring.

3   The tachometer drive pinion, made of plastic material, is retained by a circlip. Remove the circlip and the flat washer behind it, then lift off the pinion. Note that a small metal rod passes through a drilling in the end of the drive shaft, to line up with a slot in the pinion centre. When the pinion is removed, withdraw the rod and keep it in a safe place for reassembly at a later date.

4   The oil pump drive pinion is made of the same plastic material and is secured to the oil pump drive spindle by means of a nut and shakeproof washer. When the nut and washer are removed, the pinion can be pulled off the spindle.

5   It is necessary to remove the pinions only if damage has occurred. Although damage rarely occurs, the occasional chipped or broken tooth may be found when the engine is dismantled, which will necessitate renewal of the defective part.

16 Dismantling the engine unit: removing the gear selector drum

1   It is unlikely that there will be any need to remove the gear selector drum from the lower crankcase unless the tracks followed by the gear selector forks are damaged. Commence by removing the two cross-head screws that retain the gear change lever guide and the guide itself.

2   Remove the two countersunk cross-head screws that hold the retaining plate for the gear selector drum. Remove the plate.

3   Unscrew and remove the plunger assembly from the base of the crankcase. The hollow bolt that holds the plunger and spring, threads into the rear of the lower crankcase at an angle, on the left-hand side.

4   Press the gear selector drum out of its housing from the left-hand side until the circlip around the left-hand end of the drum can be removed. This will free the cam on the end of the drum, which can be pulled away. The remainder of the drum will now pass through the aperture in the right-hand end of the crankcase, permitting the drum to be removed completely.

17 Dismantling the engine unit: removing the tachometer drive

1   Removal of the tachometer drive pinion (Section 15.3) will give access to the spindle retaining plate, which is secured to the crankcase casting by three cross-head screws. Remove the three screws and the retaining plate. Pull out the drive spindle from the right, after removing the circlips which hold the worm drive pinion on the spindle.

2   The driven spindle passes through a plate attached to the top

16.2 Remove the stopper and guide plates to free drum

16.3 Unscrew the detent plunger housing

16.4 Selector cam is retained by a circlip

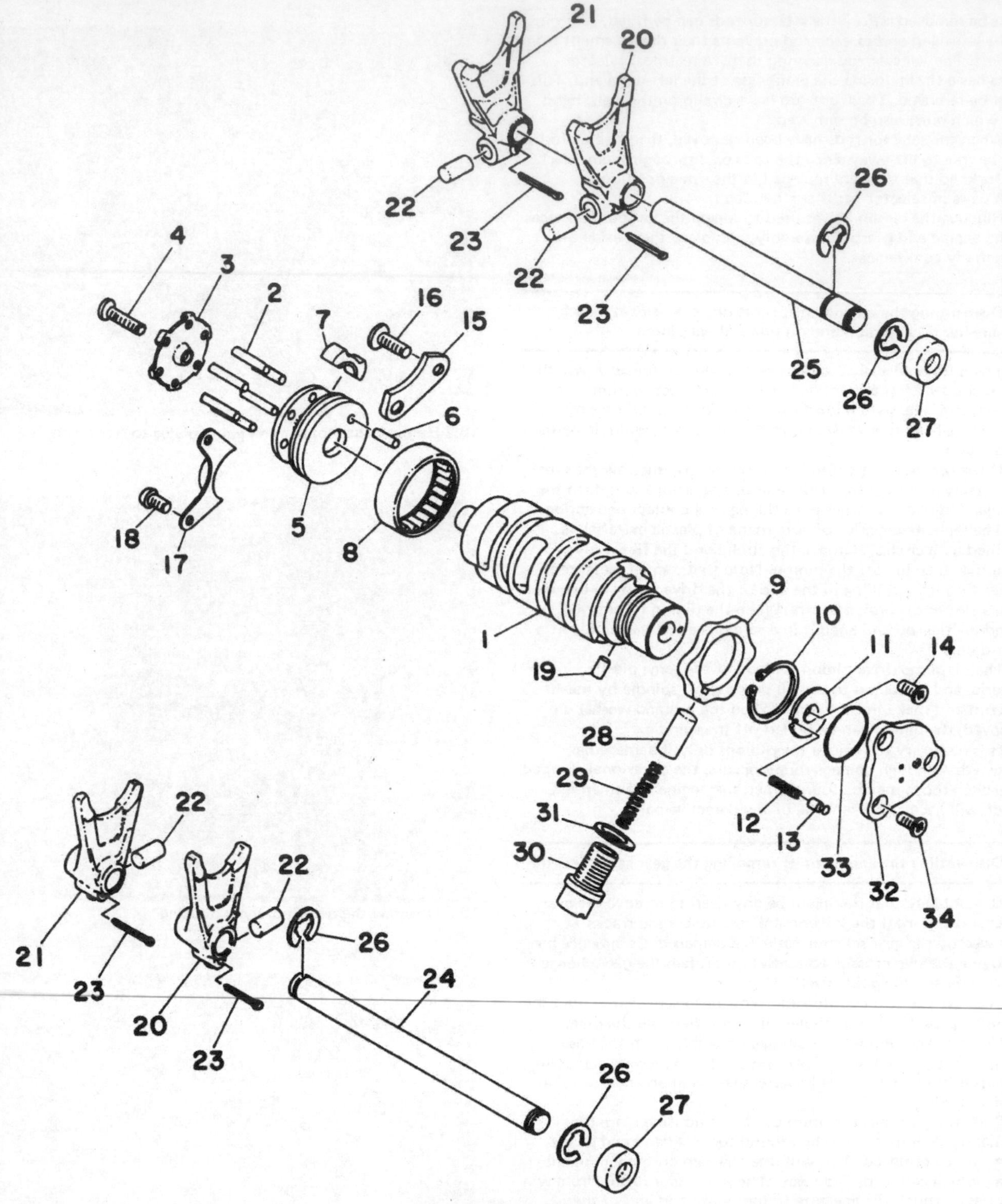

Fig. 1.6. Gearchange mechanism

1	Gearchange drum	13	Contact piece	24	Selector fork rod
2	Change pin - 6 off	14	Countersunk screw	25	Selector fork rod
3	Endplate	15	Guide plate	26	Circlip - 4 off
4	Countersunk screw	16	Countersunk - 2 off	27	Oil seal - 2 off
5	Pin carrier	17	Change lever guide	28	Detent plunger
6	Drive pin	18	Screw - 2 off	29	Detent spring
7	Pawl plate	19	Drive pin	30	Detent housing
8	Needle roller bearing	20	Selector fork A - 2 off	31	Sealing washer
9	Stopper cam	21	Selector fork B - 2 off	32	Neutral switch assembly
10	Circlip	22	Drum follower pin - 4 off	33	'O' ring
11	Neutral indicator switch plate	23	Split pin - 4 off	34	Countersunk screw - 3 off
12	Spring				

of the upper crankcase by a bolt. Remove the bolt and withdraw the driven spindle assembly complete.

18 Examination and renovation : general

1   Before examining the component parts of the dismantled engine/gear unit for wear, it is essential that they should be cleaned thoroughly. Use a paraffin/petrol mix to remove all traces of oil and sludge which may have accumulated within the engine.
2   Examine the crankcase castings for cracks or other signs of damage. If a crack is discovered, it will require professional attention, or in an extreme case, renewal of the casting.
3   Examine carefully each part to determine the extent of wear. If in doubt, check with the tolerance figures whenever they are quoted in the text. The following Sections will indicate what type of wear can be expected and in many cases, the acceptable limits.
4   Use clean, lint-free rags for cleaning and drying the various components, otherwise there is risk of small particles obstructing the internal oilways.

19 Crankshaft assembly: examination and replacement

1   The crankshaft assembly comprises two separate sets of flywheels with their respective big ends, connecting rods and main bearings, pressed together to form a single unit. A special oil seal of the labyrinth type is interposed between the two inner main bearings of each flywheel set.
2   In the event of main bearing failure, it is beyond the means of the average owner to separate the flywheel assemblies and to realign them to the high standard of accuracy required. In consequence, the complete crankshaft assembly must be taken to a Yamaha specialist for the necessary repairs and renovation, or exchanged for a fully reconditioned unit.
3   Main bearing failure will immediately be obvious when the bearings are inspected after the old oil has been washed out. If any play is evident or if the bearings do not run freely, renewal is essential. Warning of main bearing failure is usually given by a characteristic rumble that can be readily heard when the engine is running. Some vibration will also be felt, which is transmitted via the footrests.
4   Big end failure is characterised by a pronounced knock that will be most noticeable when the engine is working hard. There should be no play whatsoever in either of the connecting rods, when they are pushed and pulled in a vertical direction. A small amount of sideways play is permissible, but not more than 0.98 mm (0.038 in).
5   Oil seal failure is a common occurrence in two-stroke engines that have seen a reasonable amount of service. When the oil seals begin to wear, air is admitted to the crankcase which will dilute the incoming mixture. This in turn causes uneven running and difficulty in starting.
6   The oil seals at each end of the crankshaft are easy to renew when the engine is stripped; they are a push fit over each end of the crankshaft, one against and the other close to the outer main bearings. It is a wise precaution to renew these seals whenever the engine is stripped, irrespective of their condition.
7   The labyrinth oil seal in the centre of the crankshaft assembly can be removed and replaced only when the crankshaft assembly is separated. It is difficult to give any guide lines about the need for replacement of this seal as it normally has a long working life, akin to that of the crankshaft assembly itself. If the qualities of the middle seal are suspect, check whether the seal spins quite freely on the crankshaft. If it does, the chances are that the seal is due for renewal; it will be supplied as part of the built-up crankshaft assembly, after the Yamaha repair specialist has separated the two sets of flywheels to remove the old oil seal and re-aligned them after fitting the new seal.

17.2 Tachometer worm retained by 'E' clip

20 Small end bearings: examination and replacement

1   The small end bearings are caged needle rollers and will seldom give trouble unless lubrication failure has occurred. The gudgeon pins should be a good sliding fit in their bearings, without any play. If play develops, a noticeable rattle will be heard when the engine is running, indicative of the need for renewal of the bearings.
2   No problem is encountered when replacing the caged needle roller bearings as they are a light push fit in the eye of each connecting rod. New small end bearings are normally supplied whenever the crankshaft assembly is renewed or service-exchanged.

21 Pistons and piston rings: examination and renovation

1   Attention to the pistons and rings can be overlooked if a rebore is necessary because new pistons and rings will be fitted under these circumstances.
2   If a rebore is not considered necessary, each piston should be examined closely. Reject a piston if it is badly scored or dis-coloured as the result of the exhaust gases by-passing the rings. Check the gudgeon pin bosses to ensure that they are not enlarged or that the grooves retaining each circlip are not damaged.
3   Remove all carbon from each piston crown and use metal polish to finish off, so that a high polish is obtained. Carbon will adhere much less readily to a polished surface. Examination of the piston crown will show whether the engine has been rebored previously, since the amount of overbore is invariably stamped on each piston crown. Four oversizes are available: + 0.25 mm, + 0.50 mm, + 0.75 mm and + 1.0 mm.
4   The grooves in which the piston rings locate can become enlarged in use. The clearance between the edge of the second (lower) piston ring and the groove in which it seats should not exceed 0.05 mm (0.002 in). The top piston ring is of the Dykes type, having an 'L' shaped cross section. This type of ring is fitted to increase the ring/cylinder wall sealing by means of the pressure within the combustion chamber. The pressure forces its way behind the ring ie, between the ring groove and the inner 'L' surface of the ring thereby forcing the ring outwards. (See the accompanying diagram.)
5   Remove the piston rings by pushing the ends apart with the thumbs whilst gently easing the ring from its groove. Great care is necessary throughout this operation because the rings are brittle and will break easily if overstressed. If the rings are gummed in their grooves, three strips of tin can be used to ease

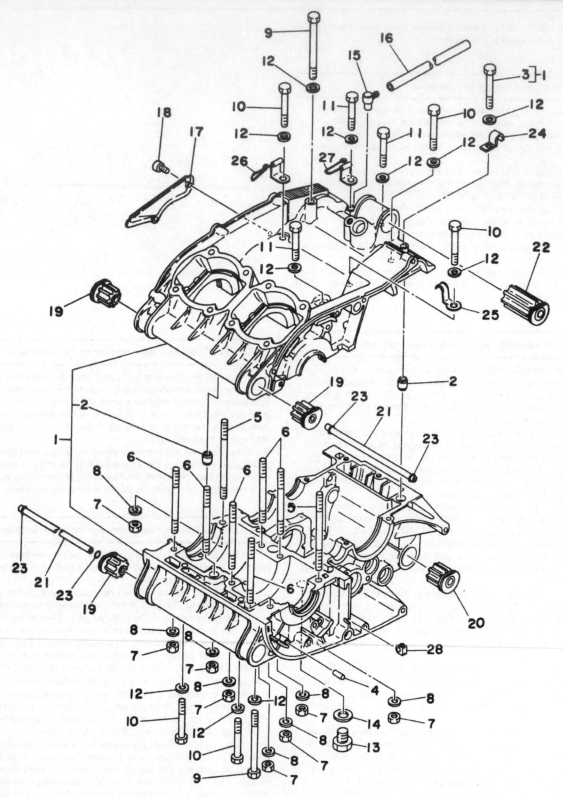

Fig. 1.7. Crankcase

1	Crankcase assembly	8	Plain washer - 8 off	15	Breather union
2	Hollow dowel - 2 off	9	Bolt - 2 off	16	Breather hose
3	Bolt	10	Bolt - 5 off	17	Oil catch plate
4	Dowel pin	11	Bolt - 3 off	18	Screw - 2 off
5	Stud - 2 off	12	Plain washer - 11 off	19	Engine mounting rubber - 4 off
6	Stud - 6 off	13	Transmission oil drain plug	20	Engine mounting rubber - 2 off
7	Nut - 8 off	14	Drain plug sealing washer	21	Engine mounting spacer - 2 off

22	Engine mounting rubber - 2 off
23	'O' ring - 4 off
24	Hose clamp
25	Hose clamp
26	Wiring clip
27	Wiring clip
28	Grommet

them free, as shown in the accompanying illustration.

6   Piston ring wear can be checked by inserting the rings one at a time in the cylinder bore from the top and pushing them down about 1½ inches with the base of the piston so that they rest square in the bore. Make sure that the end gap is away from any of the ports. If the end gap is within the range 0.3 - 0.5 mm (0.011 - 0.015 in) the ring is fit for further service.

7   Examine the working surface of each piston ring. If

Fig. 1.8. The top piston ring is of the Dykes type, having an 'L' shaped cross section

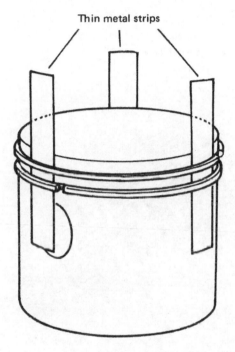

Thin metal strips

Fig. 1.9. Freeing gummed rings

discoloured areas are evident, the ring should be renewed because these areas indicate the blow-by of gas. Check that there is not a build-up of carbon on the back of the ring or in the piston ring groove, which may cause an increase in the radial pressure. A portion of broken ring affords the best means of cleaning out the piston ring grooves. The second ring is fitted with an expander band which should be checked for distortion or loss of strength. The life of an expander is similar to that of the ring, so that it is probable that the two components will not require replacing independently.

8   Check that the piston ring pegs are firmly embedded in each piston ring groove. It is imperative that these retainers should not work loose, otherwise the rings will be free to rotate and there is danger of the ends being trapped in the ports.

9   It cannot be over-emphasised that the condition of the pistons and piston rings is of prime importance because they control the opening and closing of the ports by providing an effective moving seal. A two-stroke engine has only three working parts, of which the piston is one. It follows that the efficiency of the engine is very dependent on the condition of pistons and the parts with which they are closely associated.

22 Cylinder barrels: examination and renovation

1   There will probably be a lip at the uppermost end of each cylinder barrel which marks the limit of travel of the top of the piston ring. The depth of the lip will give some indication of the amount of bore wear that has taken place even though the amount of wear is not evenly distributed.

2   Insert each piston (without rings) in turn in its respective cylinder bore so that it is about ¾ in from the top of the bore. Measure the clearance between the piston skirt and the cylinder wall with a feeler gauge. Repeat at two further positions lower down the bore. The recommended minimum clearance is 0.035 mm (0.0013 in). If the clearance exceeds 0.040 mm (0.0016 in) the cylinder is in need of a rebore. Always rebore the cylinders as a pair and fit oversize pistons.

3   Give the cylinder barrels a close visual inspection. If the surface of either bore is scored or grooved, indicative of an earlier seizure or a displaced circlip and/or gudgeon pin, rebore is essential. Compression loss has a marked effect on engine performance.

4   Check that the outside of each cylinder barrel is clean and free from road dirt. If the air flow to the cooling fins is obstructed, the engine may overheat badly. Although caustic soda is often recommended for cleaning some of the oilier components of a two-stroke engine, NEVER use on parts made of light alloy.

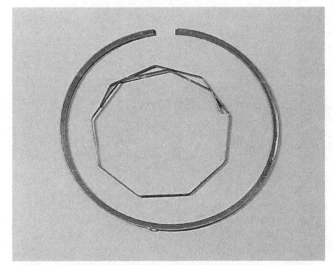

21.7 An expander ring is fitted inside the second ring on each piston

21.8 Check that the ring pegs are not loose

Caustic soda attacks aluminium alloy with great vigour and produces an explosive gas.

5   Clean all carbon deposits from the exhaust ports, using a blunt ended scraper. It is important that all the ports should have a clean, smooth appearance because this will have the dual benefit of improving gas flow and making it less easy for carbon to adhere in the future. Finish off with metal polish, to heighten the polishing effect.

6   Do not under any circumstances enlarge or alter the shape of the ports, under the mistaken belief that improved performance will result. The size and position of the ports predetermines the characteristics of the engine and unwarranted tampering can produce very adverse effects.

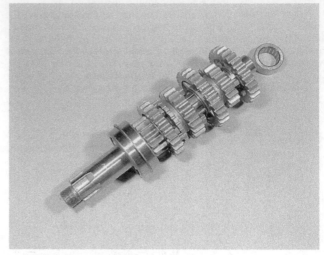

24.1a Mainshaft - general view

23 Cylinder heads: examination and renovation

1   It is unlikely that the cylinder heads will require any special attention apart from removing the carbon deposit from the combustion chamber. Finish off with metal polish; the polished surface will help improve fast flow and reduce the tendency of future carbon deposits to adhere so readily.

2   Check that the cooling fins are clean and unobstructed, so that they receive the full air flow. Use of a wire brush will damage the black finish on the cylinder heads.

3   Check the condition of the thread within each spark plug hole. The thread is easily damaged if the spark plug is over-tightened. If necessary, a damaged thread can be reclaimed by fitting a Helicoil thread insert. Most Yamaha agents have facilities for this type of repair, which is not expensive.

4   If there has been evidence of oil seepage from the cylinder head joint when the machine was in use, check whether either cylinder head is distorted by laying it on a sheet of plate glass. Severe distortion will necessitate renewal of the cylinder head, but if the distortion is only slight, the head can be reclaimed by wrapping a sheet of emery cloth around the glass and using it as the surface on which to rub down the head with a rotary motion, until it is once again flat. The usual cause of distortion is failure to tighten down the cylinder head bolts evenly in a diagonal sequence.

24.1b Layshaft - general view

24 Gearbox components: examination and renovation

1   Examine each of the gear pinions to ensure that there are no chipped or broken teeth and that the dogs on the end of the pinions are not rounded. Gear pinions with any of these defects must be renewed; there is no satisfactory method of reclaiming them.

2   The gearbox bearings must be free from play and show no signs of roughness when they are rotated. Each shaft has a ball journal bearing at one end and a caged needle roller bearing at the other.

3   It is advisable to renew the gearbox oil seals irrespective of their condition. Should a re-used oil seal fail at a later date, a considerable amount of dismantling is necessary to gain access and renew it.

4   Check the gear selector rods for straightness by rolling them on a sheet of plate glass. A bent rod will cause difficulty in selecting gears and will make the gear change action particularly heavy.

5   The selector forks should be examined closely, to ensure that they are not bent or badly worn. Wear is unlikely to occur unless the gearbox has been run for a period with a particularly low oil content.

6   The tracks in the gear selector drum, with which the selector forks engage, should not show any undue signs of wear unless neglect has led to under lubrication of the gearbox. Check that the plunger spring bearing on the cam plate plunger has not lost its action and that the springs of the gear change lever pawl assembly have good tension. Any damage to, or weakness of, the gear change lever return spring will be self-evident.

7   If the kickstart has shown a tendency to slip, or if it is

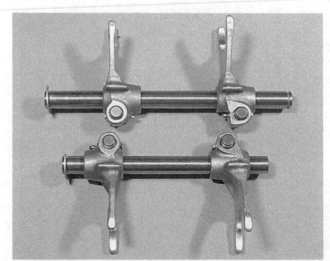

24.5 Check for wear on the fork ends and the pins

desired to inspect the kickstart ratchet, the kickstart assembly must be dismantled. Remove the circlip from the splined end of the kickstart shaft and then withdraw the spacer, spring cover and spring clip around the ratchet wheel. Any wear of the ratchet teeth will be self-evident and it will be necessary to renew both the ratchet wheel and the kickstart gear pinion with which it makes contact. Examine the kickstart return spring, which should also be renewed if there is any doubt about its condition.

8 As mentioned earlier in the text, instances have occurred where a tooth has either chipped or broken on the pinions that drive the oil pump and the tachometer drive cable take-off. These pinions are made of plastic material and must be renewed when damage of this nature is found.

25 Clutch assembly: examination and renovation

1 After an extended period of service, the clutch linings will wear and promote clutch slip. The limit of wear measured across each inserted plate is 0.27 mm (0.107 in). When the overall width of the linings reach this level, the clutch plates must be replaced as a complete set.

2 The plain clutch plates should not show any evidence of excess heating (blueing) and should not be more than 0.05 mm (0.002 in) out of true.

3 The clutch springs should have a free (uncompressed) length of 36.4 mm. If the springs have taken a set of 1 mm or more, the complete set must be renewed.

4 A worn clutch spacer is responsible for clutch noise and should be renewed if the fit within the clutch centre is particularly slack. Check the inner and outer surfaces for scratches; these will impair clutch action if not smoothed away.

5 Check the condition of the slots in the outer surface of the clutch centre and the inner surfaces of the outer drum. In an extreme case, clutch chatter may have caused the tongues of the inserted plates to make indentions in the slots of the outer drum, or the tongues of the plain plates to indent the slots of the clutch centre. These indentations will trap the clutch plates as they are freed and impair clutch action. If the damage is only slight the indentations can be removed by careful work with a file and the burrs removed from the tongues of the clutch plates in similar fashion. More extensive damage will necessitate renewal of the parts concerned.

6 The clutch release mechanism attached to the inside of the left-hand crankcase cover does not normally require attention, provided it is greased at regular intervals. It is held to the cover

24.6 Check for wear in the change drum channels

24.7a Kickstart spring is retained by a collar ...

24.7b ... held on the shaft by a circlip

24.7c Check the pawl and ratchet teeth

by two cross-head screws and operates on the worm and quick
start thread principle. A light return spring ensures that the
pressure is taken from the end of the clutch push rod when the
handlebar lever is released and the clutch fully engaged.

26 Crankcase covers: examination and renovation

1 The right and left-hand crankcase covers are unlikely to
suffer damage unless the machine is dropped or damaged in an
accident. If the right-hand cover is fractured, the kickstart will
be rendered inoperative since the cover acts as the outer bearing
for the kickstart shaft. Renewal is therefore essential.
2 The covers have a matt black anodised finish, accentuated by
raised portions of the casting that have a polished surface. Small
chips can be repaired by over-painting with a matt finish such as
cylinder black. This is only a temporary expedient. To fully
restore the original finish it will be necessary to have the outer
surface of the cover shotblasted and then re-anodised, or
alternatively to renew the cover.

27 Engine reassembly: general

1 Before reassembly of the engine/gear unit is commenced, the
various component parts should be cleaned thoroughly and
placed on a sheet of clean paper, close to the working area.
2 Make sure all traces of old gaskets have been removed and
that the mating surfaces are clean and undamaged. One of the
best ways to remove old gasket cement is to apply a rag soaked
in methylated spirit. This acts as a solvent and will ensure
that the cement is removed without resort to scraping and the
consequent risk of damage.
3 Gather together all the necessary tools and have available an
oil can filled with clean engine oil. Make sure all new gaskets
and oil seals are to hand, also all replacement parts required.
Nothing is more frustrating than having to stop in the middle of
a reassembly sequence because a vital gasket or replacement has
been overlooked.
4 Make sure that the reassembly area is clean and that there is
adequate working space. Refer to the torque and clearance
settings wherever they are given. Many of the smaller bolts are
easily sheared if over-tightened. Always use the correct sized
screwdriver bit for the cross-head screws and never an ordinary
screwdriver or punch. If the existing screws show evidence of
maltreatment in the past, it is advisable to renew them as a
complete set.

28 Engine reassembly: replacing the gear selector mechanism

1 Engine assembly commences with the lower crankcase, which
should be positioned in the centre of the workbench. If the gear
selector drum has been removed, this must be replaced first by
reversing the dismantling procedure given in Section 16. It is
important that the circlip retaining the camplate at the end of
the drum is located in the correct position. The accompanying
illustration shows how this must be aligned.
2 Insert the gear selector rods from the left-hand side of the
engine, one at a time. There are two types of selector fork
fitted, one each of which type is fitted on each selector fork.
Consult the accompanying photograph for correct fork position-
ing.
3 The pins that engage with the selector drum tracks are a push
fit in the selector forks and are easily replaced.
4 There is no necessity to select any particular gear during
assembly of the gearbox since the system of gear index is
automatic and does not require 'timing'.
5 When the selector rods and forks are in position, fit new oil
seals into the recesses on the left-hand side of the crankcase. They
serve the dual purpose of retaining the selector rods in position
after the locating circlips.
6 Replace the camplate plunger assembly in the underside of

28.1a Insert change drum through gearbox wall and ...

28.1b ... refit the selector cam and circlip

28.1c Replace the change drum guide plate and ...

28.1d ... the stopper plate, both retained by two screws

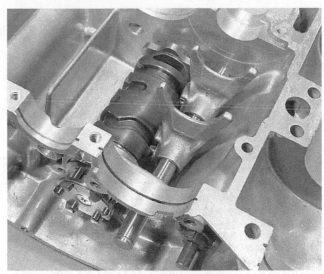
28.2a Replace both selector fork sets ...

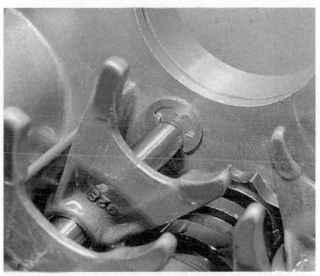
28.2b DO NOT omit the 'E' clips

28.2c Selector fork positions - general view

28.2d Drift the blind seals into the selector rod registers

28.6 Refit the camplate detent unit into the crankcase

the lower crankcase and check that the end cap is tight and has a new sealing washer.

7 Position the lower crankcase with the right-hand side uppermost and refit the gear selector mechanism. The operating arm fits over a spindle which protrudes from the crankcase casting; it is retained in position by a circlip. Lubricate the spindle before the operating arm is replaced. It will be necessary to hold apart the two spring-loaded pawls whilst the arm is lowered into position, so that they clear the pins on the end of the gear selector drum.

29 Engine reassembly: replacing the gear clusters

1 Engine reassembly should only be commenced after any dismantled gear train has been reassembled as a complete unit with the gearbox bearings.

2 Position the lower crankcase on its base and insert the three bearing retainers that take the form of half circlips. Lower each complete gear cluster in turn so that the bearings engage with the circlips and the selector fork(s) with the track around the sliding gear pinion. It may be necessary to tap each gear shaft with a soft-faced mallet before they will bed down completely. Check that the shafts revolve quite freely; if not, some part

of the assembly is out of alignment.

3 Oil the shafts, gear pinions and bearings so that there will be no dry spots during the eventual start-up of the engine.

30 Engine reassembly: replacing the crankshaft

1 Replace the half circlips bearing retainer in the right-hand bearing housing. Lower the crankshaft assembly into position, noting that each of the main bearings has what is known as a 'knock pin' which engages with a small depression at the leading edge of the crankcase joint.

2 Give both ends of the crankshaft assembly a light tap with a soft-faced mallet, to ensure the assembly is located correctly. Check that each of the knock pins is seating correctly in the bearing housings. Position the crankcase locating dowels in the front and rear of the lower crankcase casting.

31 Engine reassembly: joining the upper and lower crankcase

1 Check that the upper crankcase casting has the tachometer drive assembly fitted. If it is necessary to replace the drive components, reverse the dismantling procedure described in

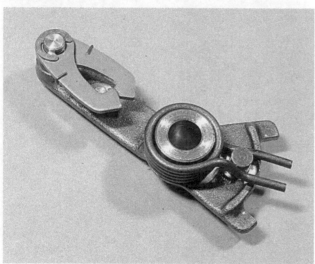

28.7a Ensure the gearchange pawl arm spring is fitted as shown

28.7b Fit pawl arm with spring 'ears' either side of anchor

29.2a Ensure that the bearing half-clips are correctly positioned

29.2b ... replace the gearshaft assemblies one at a time ...

29.2c ... ensuring that the bearings locate with the half-clips

29.2d Ensure that the seals are positioned correctly

30.1a Position the special 'O' ring on the crankshaft

30.1b Lower the crankshaft into position

30.1c Engage the 'knock' pins with the recesses

30.2a The right-hand crankshaft seal engages with a recess ...

30.2b ... as does the left-hand seal. Both must be positioned correctly

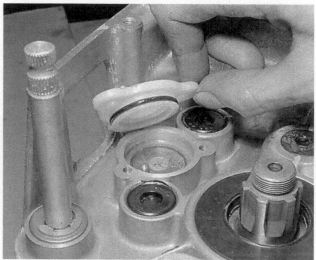

32.2 DO NOT omit 'O' ring on neutral switch cover

34.2 Lubricate final drive sprocket collar before fitting

Section 17.
2 Coat the jointing surfaces of both crankcases with a thin layer of gasket cement and lower the upper crankcase casting into position. It may be necessary to give the casting a few light taps with a soft-faced mallet before the jointing surfaces will mate up correctly. DO NOT USE FORCE. If the crankcase will not align, one of the bearing assemblies has not seated correctly.
3 Replace the eight bolts in the top of the crankcase and tighten them in the numerical sequence stamped on the crankcase casting.
4 Invert the crankcase and replace the eight nuts and three bolts on the underside. These too should be tightened in the numerical sequence indicated.

32 Engine reassembly: replacing the neutral indicator

1 If the neutral indicator has been removed from the left-hand end of the gear selector drum it will be necessary first to replace the spring and metal contact, then the circular plate attached to the end of the drum by means of a central, cross-head screw. Before the circular plate is tightened down, check that the spring-loaded contact is engaged correctly with the slot in the plate and will move quite freely. The metal contact is waisted to engage with the slot.
2 Replace the end cover; secure by the three countersunk cross-head screws. The end cover is made of a plastics material and has an inner 'O' ring seal, which must be in good condition.

33 Engine reassembly: replacing the crankshaft and gearbox oil seals

1 The crankshaft assembly has two oil seals, one of large diameter on the right-hand side and the other of smaller diameter on the left-hand side. The smaller of the two seals is replaced in what may at first appear to be the reverse position, with the spring around the centre boss facing OUTWARD. As a check, the face of the seal is marked to show which is the outer side. The larger seal is fitted in the conventional manner and should have the serrated portion of the face positioned inward, towards the main bearing. The left-hand oil seal should be placed so that the outer face is flush with the face of the crankcase boss. The right-hand seal should be fitted so that the inner face is against the outer face of the main bearing. The crankshaft oil seals must be refitted before the crankcases are rejoined.
2 The gearbox has three oil seals, two on the left-hand side and one on the right. There is a large diameter seal immediately in front of both the main bearings and a smaller seal in front of the end of the hollow mainshaft, through which the clutch push rod protrudes.
3 Before the oil seals are replaced, the shafts over which they fit should be lubricated thoroughly. Take care to ensure the lips of the seals are not damaged as they are driven into position. On no account use force, otherwise the risk of damage will be high. When positioned correctly, the outer face of each seal should be flush with the edge of the crankcase casting. Before continuing with the reassembly, check that the crankshaft and the gearbox shafts still revolve quite freely.

34 Engine reassembly: replacing the final drive sprocket

1 With the engine resting with its left-hand side uppermost, it is convenient to replace the final drive sprocket. If the original is badly worn or has hooked or broken teeth, it must be renewed at this stage, otherwise a harsh transmission and very rapid chain wear will occur.
2 The sprocket fits over the splined end of the gearbox layshaft and is retained by a large diameter nut and tab washer which also fits over the splines. Lubricate and refit the sprocket distance collar before replacing the sprocket.
3 Hold the sprocket firmly by wrapping the chain around the

teeth and securing both ends in a vice, then tighten the centre nut fully (right-hand thread). Bend the tab washer so that it locks the retaining nut in position.

35 Engine reassembly: replacing the primary drive pinion

1   The primary drive pinion is retained on the parallel-shouldered end of the crankshaft by an oblong key and a keyway. Oil the outer surface of the shoulder behind the pinion teeth and slide the pinion on to the crankshaft so that the keyways in both shaft and pinion coincide. Drive the oblong key into position, then add the domed washer and the pinion

retaining nut. Lock the crankshaft prior to tightening the nut, by means of a close fitting bar through the small end eye of one of the connecting rods.

36 Engine reassembly: replacing the kickstart mechanism

1   If the kickstart shaft assembly has been dismantled, it must be reassembled as a complete unit, prior to replacement in the outer crankcase. The accompanying diagram shows the correct sequence of assembly.
2   Lubricate the shaft before it is inserted into the crankcase boss and arrange the stop so that it is hard against the crankcase

34.3 Remember to bend up tab washer

35.1 Primary drive pinion must engage with 'square' key

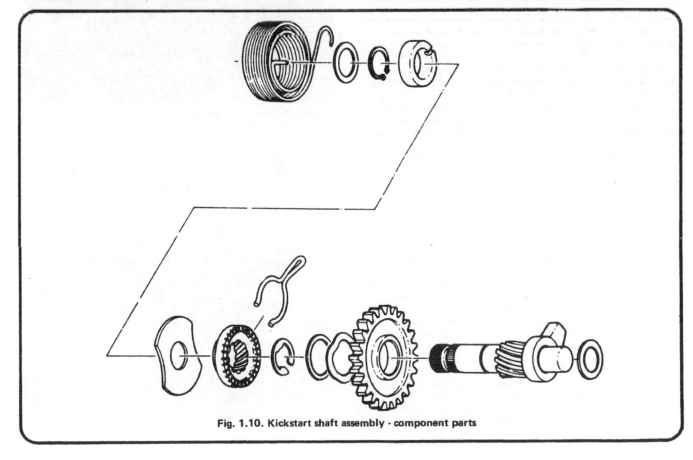

Fig. 1.10. Kickstart shaft assembly - component parts

abutment when the shaft is turned in a clockwise direction.
Under these circumstances there will be a positive tension on the
kickstart return spring, even when the kickstart is in the
fully-raised position. The end of the spring locates with a peg
projection from the casing and should be turned anti-clockwise
to locate without moving the shaft. The ratchet wheel locates
with a cutaway below this peg.

37 Engine reassembly: replacing the tachometer drive pinion

1   Although there is no necessity to remove the drive pinion
under normal circumstances (it does not impede dismantling or
reassembly) it may be necessary to replace the pinion on the odd
occasion.
2   A short metal rod passes through the spindle on which the
pinion runs, which should be inserted and spaced so that both
ends are equally spaced from the centre. Lower the pinion into
position so that the ends of the rod engage with the slots in the
centre boss. Replace the washer which fits within the boss, then
the circlip that secures the whole assembly.

38 Engine reassembly: replacing the gear change shaft

1   Insert a new oil seal into the lower passageway on the
left-hand side of the engine, through which the gear change shaft
passes.
2   Insert the gear change shaft from the right-hand side,
lubricating the shaft to ease its passage through the oil seal.
3   Slip the shouldered collar, larger diameter downward, over
the peg on the gear selector arm and press home the gear change
shaft so that the slotted portion of the operating arm engages
with the shouldered collar and peg.
4   Hold the gear change shaft in this position and replace the
shim washer and circlip around the left-hand end of the shaft, to
retain it in position.

39 Engine reassembly: reassembling and replacing the clutch

1   Position the engine so that the right-hand side faces uppermost and replace the thrust washer over the gearbox mainshaft,
immediately in front of the main bearing.
2   Replace the kickstart idler pinion over the shaft adjacent to
the mainshaft because this cannot be replaced when the main
body of the clutch is in position. It has a Belville washer and a
plain washer below it, and is retained by a circlip.
3   Slide the spacer over the gearbox mainshaft and then lower
the clutch outer drum and integral primary drive gear into
position so that it meshes with both the primary drive pinion
and the kickstart idler.
4   Position a further plain washer over the mainshaft which
protrudes through the clutch outer drum, then position the
clutch centre to engage with the splines on the end of the
mainshaft. Replace the domed spring washer and the centre nut
retaining the clutch centre in position (right-hand thread).
5   Lock the gearbox in position by wrapping chain around the
final drive sprocket and holding both ends in a vice. The centre
retaining nut of the clutch assembly can now be tightened.
Release the chain from the vice and remove it from the sprocket.
6   The clutch plates can now be inserted one at a time,
commencing with a friction plate. Note that a rubber ring is
interposed between each following set of friction and plain
plates, so that the plain plates cannot come into contact with
one another when the clutch is withdrawn and create clutch
noise. The sequence of assembly is, therefore, friction plate,
rubber ring, plain plate, friction plate, rubber ring etc. Before the
pressure plate is positioned, make sure the clutch push rod
'mushroom' has been greased and inserted into the hollow end of
the mainshaft.
7   Replace the six clutch springs in the clutch pressure plate and

36.2a Refit kickstart assembly. Note lower shim and ...

36.2b ... the two upper shims

36.2c Tension return spring. Ensure clip spring locates in recess

38.2 Insert change shaft to mesh with pawl arm

39.2a Refit wave type shim on shaft and ...

39.2b ... replace the kickstart idler pinion and circlip

39.3a Lubricate and fit the clutch spacer and washer

39.3b Refit the clutch outer drum and washer and ...

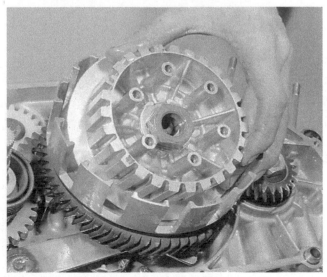
39.4 ... the clutch centre, nut and tab washer

39.6a Commence insertion of plates with a friction plate ...

39.6b ... followed by a plain plate and expander ring

39.6c Insert the 'mushroom' head clutch piece and ...

39.7 ... refit the pressure plate and clutch springs

40.3a Position each piston so that the arrow faces forward

40.3b Fit the gudgeon pin. ALWAYS use new circlips

tighten down the retaining screws until the full depth of thread is engaged. The pressure plate must be fitted so that the arrow mark on the periphery aligns with one of the three arrows on the centre boss.

40 Engine reassembly: replacing the pistons and piston rings

1   Position the engine to rest on the base of the crankcase. Pad the mouth of each crankcase with clean rag prior to fitting the pistons and piston rings, so that any displaced parts will be prevented from falling in.
2   Replace the caged needle roller bearings in each small end, then fit the pistons and gudgeon pins, checking to ensure that they are replaced in their original locations. Note that the pistons have an arrow stamped on the crown, which must face forwards.
3   If the gudgeon pins are a tight fit in the piston bosses, the pistons can be warmed with warm water to effect the necessary temporary expansion. Oil the gudgeon pins and piston bosses before the gudgeon pins are inserted, then fit the circlips, making sure that they are engaged fully with their retaining grooves. A good fit is essential, since a displaced circlip will cause extensive engine damage. Always fit new circlips, NEVER re-use the old ones.
4   Check that the piston rings are fitted correctly, with their ends either side of the ring pegs and the expanders behind the lower rings. If this precaution is not observed, the rings will be broken during assembly.

41 Engine reassembly: replacing the cylinder barrels

1   Place a new cylinder base gasket over the retaining studs and lubricate each cylinder bore with clean engine oil. Arrange the left-hand piston so that it is at top dead centre (TDC) and lower the left-hand cylinder down the retaining studs until contact is made with the piston. The rings can now be squeezed one at a time until the cylinder barrel will slide over them, checking to ensure that the ends are still each side of the ring peg. Great care is necessary during this operation, since the rings are brittle and very easily broken.
2   Although the cylinder barrels have a good lead-in, to facilitate entry of the piston rings, a piston ring clamp can be used as an alternative to the hand feed method. Here again, care must be taken to ensure that the rings are correctly positioned in relation to the piston ring pegs.
3   When the rings have engaged fully with the cylinder bore, withdraw the rag packing from the crankcase mouth and slide

the cylinder barrel down the retaining studs, so that it seats on the new base gasket (no gasket cement).
4   Repeat this procedure for the right-hand cylinder, after arranging the right-hand piston so that it is at TDC.

42 Engine reassembly: replacing the cylinder heads

1   Because the cylinder heads have a full hemisphere and have central spark plugs, they are identical and fully interchangeable.
2   Place a new copper cylinder head gasket on the top of each cylinder barrel, using a smear of grease to retain them in position. Fit each cylinder head in turn, taking care that the cylinder head gasket is not displaced or distorted during the initial tightening down.
3   Each cylinder head has four sleeve bolts, which must be tightened evenly, in a diagonal sequence. This is most important since distortion will occur if this precaution is not observed. Use a torque wrench to achieve the final setting of 15 lb-ft (2 kg-m).
4   Replace the spark plugs in order to prevent any extraneous material from dropping into the engine, especially whilst it is being refitted into the frame.

43 Engine reassembly: replacing the right-hand crankcase cover and oil pump

1   The oil pump is attached to the right-hand crankcase cover and need not be disturbed unless the cover is broken, necessitating renewal. If it is necessary to transfer the pump from one case to another, follow the procedure given in Chapter 2, Section 13.
2   Since the outer crankcase cover contains the gearbox oil filler and has to retain the oil content, a good joint between the outer crankcase and the cover is essential. A gasket is used at the jointing face; provided the jointing surfaces are in good condition and undistorted, the joint can be made dry. If there is any doubt, give both surfaces a light coating of gasket cement.
3   The cover is retained by nine cross-head screws. Care is necessary to ensure the oil pump driving pinion meshes with the primary drive pinion; if the cover is forced into position, teeth may be broken from the plastic drive pinion. The cover has two dowels to aid location.

44 Engine reassembly: replacing the alternator

1   Provided care is exercised when replacing the engine unit in

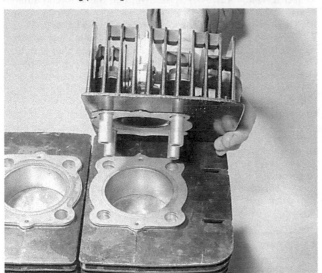

42.3 Fit new head gasket and replace cylinder heads

43.2 Use a new gasket at the primary drive cover joint

the frame, there is no reason why the alternator should not be fitted at this stage. Replace the woodruff key in the tapered left-hand end of the crankshaft and position the rotor on the shaft so that the keyways align. The rotor is retained by a centre bolt passed through the centre of the ignition cam. The latter is pegged, to ensure it is located correctly.

2   Tighten the centre bolt, then replace the contact breaker assembly and brush gear cover, which is attached by three long cross-head screws. The wiring harness feeds through an aperture in the rear of the upper crankcase casting and is sealed by a rubber grommet. A clip, through which one of the upper crankcase bolts passes, acts as a cable retainer. A small dowel ensures that the cover can be replaced in one position only.

3   Reconnect the short lead to the neutral indicator switch by the spade end connector and join up the main harness with the snap connector provided. A small, square-shaped rubber grommet seals the gap in one of the crankcase ribs, through which the wire passes. The engine unit is now ready for refitting in the frame.

45 Refitting the engine/gear unit in the frame

1   Place the machine on the centre stand, so that it is standing rigidly on firm ground. Lift the engine unit and with the aid of an assistant, slide the engine unit into the frame from the right-hand side. The front engine mountings are waisted, to correspond with the design of the front of the crankcase, so that there is sufficient clearance for insertion.

2   When the engine unit is in approximately the correct position, fit the two front engine bolts, and the engine bolt close to the swinging arm pivot. Do not tighten the bolts at this stage.

3   Replace the short upper rear engine bolt which passes through a lug cast into the crankcase and two small engine plates. Replace the two bolts that retain these upper engine plates, then tighten all the bolts to the recommended torque settings of 2 kg-m (15 lb-ft) for the 8 mm bolts and 3.5 kg-m (25 lb-ft) for the 10 mm bolts.

46 Engine reassembly: reconnecting the oil pipes and carburettors

1   Reconnect the main oil feed pipe to the oil pump inlet after threading the pipe through the rubber grommet at the top of the right-hand crankcase cover. The pipe is held in position by a small wire clip.

2   Fit the twin carburettors to the flexible induction stubs. Note that the two carburettors are inter-connected by a short rubber balance pipe, and that the carburettor with the choke lever is fitted on the left. Each carburettor is retained by a metal clip around the flexible rubber intake into which the carburettor body pushes. Reconnect the flexible oil pipes to the unions on the carburettors, ensuring that the securing spring clips are correctly placed.

3   Replace the carburettor tops, complete with the throttle slide and needle assembly. The slot in the side of each slide must locate with a projection within the mixing chamber, before the slide can be lowered into its correct position and the top screwed home. Before reconnecting the carburettor intakes with the air cleaner assembly, check that the slides are synchronised correctly, as described in Chapter 2, Section 8. A short rubber hose connects each intake with the air cleaner; it is a push on fit over the air cleaner joint and is secured to the carburettor intake and air box by means of metal screw clips.

5   Group the carburettor drain pipes without sharp bends so that they will vent freely.

47 Engine reassembly: reconnecting, bleeding and setting the oil pump

1   Reconnect the wire control cable linking the oil pump with the twist grip throttle. It is best to arrange the wire in a

loop and engage the nipple with the oil pump pulley, before the cable is seated in the pulley groove.

2   Because the oil feed pipe now contains air, it is necessary to bleed the oil pump until all the air bubbles are removed. Check that the oil tank is filled with oil, then unscrew and remove the small cross-head screw, which has a fibre washer beneath the head, from the oil pump body. Rotate the plastic wheel with the milled edge at the rear of the oil pump, in a CLOCKWISE direction (as denoted by the arrow marking) and continue turning until the oil commences to flow from the outlet which the bleed screw normally seals. Continue turning until all air bubbles have been eliminated from the main feed, then replace the screw and washer. On some later models the plastic wheel is not fitted. Bleeding may be carried out by operating the twist grip; this will rotate the pulley and so activate the pump plunger. Open and shut the twist grip until bleeding is accomplished.

3   To check whether the pump opening is correct, follow the procedure given in Chapter 2, Section 15. If the pump was set up correctly initially, it is improbable that significant changes in setting will be required. DO NOT OMIT THIS CHECK UNDER THE ASSUMPTION IT MUST BE CORRECT.

4   Replace the semi-circular cover over the oil pump, which is retained by three cross-head screws.

48 Engine reassembly: replacing the left-hand crankcase cover

1   Before replacing the left-hand crankcase cover, check that the contact breaker gap and ignition timing is correct for both cylinders. An aperture in the generator cover has a pointer which will align with timing marks on the rotor making checking of the ignition timing quite simple. See Chapter 3, Section 9 for details. Loop the final drive chain over the gearbox sprocket.

2   Replace the clutch cable into the small portion of crankcase lip which acts as a cable stop in the top of the outer crankcase housing. Reconnect the cable nipple with the clutch release mechanism in the inside of the left-hand crankcase cover and refit the cover; this is retained by three cross-head screws. To make cable fitting easy, slacken off the adjuster at the handlebar lever, or even remove the cable from the lever itself. Before the cover is replaced, do not omit to replace the clutch push rod, which is preceded by a ball bearing, and the plastic cover over the gear change lever shaft. The push rod should be well greased, prior to insertion. The ball bearing at the other (left-hand) end is best retained within the clutch operating mechanism of the left-hand crankcase cover by a dab of grease.

3   Refit the circular generator cover, and secure with the three long cross-head screws which pass through the left-hand outer cover.

4   There is no gasket at the left-hand crankcase cover joint.

5   Before fitting the small insert in the left-hand cover which masks off the clutch operating mechanism, check that clutch adjustment is correct. There should be a small amount of play when the adjusting screw is turned in a clockwise direction, before pressure is applied to the clutch push rod. If necessary, fit the clutch cable to the handlebar lever adjuster, adjust the operating mechanism so that there is about 3/32 in play at the handlebar lever before the clutch action commences. Adjustment is effected by slackening the locknut of the adjusting screw and turning the screw clockwise to decrease the clearance or anti-clockwise to increase it, before tightening the locknut and re-checking.

6   Reconnect the final drive chain. It is easiest to fit the spring link when both ends of the chain are pressed into the rear wheel sprocket.

49 Engine reassembly: replacing the exhaust pipes

1   Fit the nylon insert into each silencer front pipe and replace the rubber connectors. Insert the exhaust pipes and push them home fully. Position a new exhaust gasket in each exhaust port and align the mating faces of the exhaust pipes with the gasket.

44.1a Replace the Woodruff key in the crankshaft and ...

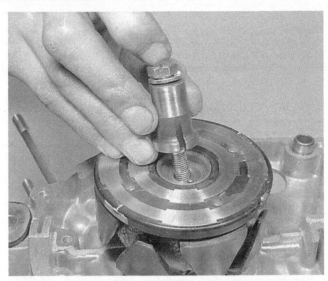

44.1b ...fit alternator. The cam engages with peg in alternator

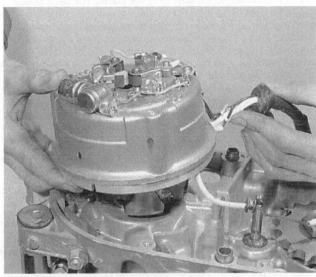

44.2a Carefully fit the alternator stator noting that ...

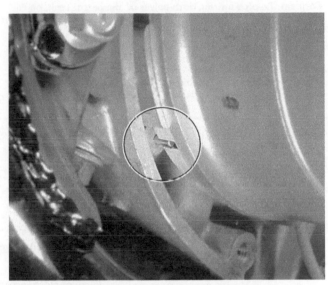

44.2b ... the stator must engage with peg in casing

48.2a Insert ball bearing and clutch pushrod

48.2b Connect cable and grease clutch lifter

Position the flanges and tighten the nuts down evenly, until it is evident that the gaskets are compressed and a good seal is made. Do not overtighten the flange nuts.

50 Engine reassembly: completion and final adjustments

1   Remake the electrical connections from the alternator harness. The various connectors should be protected by the plastic sleeve which surrounds them, close to the right-hand carburettor. A clip attached to one of the top, left-hand crankcase bolts acts as a guide for the wiring harness, and prevents the harness from chafing or rubbing. Reconnect the battery and the spark plug leads, and check that the lights and ignition circuit are live when switched on.

2   Replace the kickstart and the gear change lever. Both fit on splines and are retained by a pinch bolt. Check that they are at the desired operating angle before tightening the pinch bolt.

3   Replace the petrol tank, making sure that the mounting rubbers engage with the retainers close to the steering head. Connect the pipe uniting both halves of the petrol tank, on the underside, and make sure it is retained by wire clips. Connect the feed pipes to the carburettors and turn the petrol tap to the off position. Refill the tank with petrol.

4   Check the contents of the oil tank and top up if necessary with oil SAE 30 two-stroke grade. Remove the filler cap from the right-hand crankcase cover and add 1700 cc of SAE IOW/30 engine oil. Check with the dipstick that the level is correct.

51 Starting and running the rebuilt engine

1   When the initial start-up is made, run the engine slowly for the first few minutes, especially if the engine has been rebored or a new crankshaft fitted. Check that all the controls function correctly and that there are no oil leaks before taking the machine on the road. The exhausts will emit a high proportion of white smoke during the first few miles, as the excess oil used whilst the engine was reassembled is burnt away. The volume of smoke should gradually diminish until only the customary light blue haze is observed during normal running. It is wise to carry a spare pair of spark plugs during the first run, since the existing plugs may oil up due to the temporary excess of oil.

2   Remember that a good seal between the pistons and the cylinder barrels is essential for the correct functioning of the engine. A rebored two-stroke engine will require more carefully running-in, over a longer period, than its four-stroke counterpart. There is a far greater risk of engine seizure during the first hundred miles if the engine is permitted to work hard.

3   Do not tamper with the exhaust system or run the engine without the baffles fitted to the silencer. Unwarranted changes in the exhaust system will have a very marked effect on engine performance, invariably for the worse. The same advice applies to dispensing with the air cleaner or the air cleaner element.

4   Do not on any account add oil to the petrol under the mistaken belief that a little extra oil will improve the engine lubrication. Apart from creating excess smoke, the addition of oil will make the mixture much weaker, with the consequent risk of overheating and engine seizure. The oil pump alone should provide full engine lubrication.

52 Fault diagnosis: engine

Symptom	Cause	Remedy
Engine will not start	Defective spark plugs	Remove plugs and lay on cylinder head. Check whether spark occurs when engine is kicked over.
	Dirty or closed contact breaker points	Check condition of points and whether gap is correct.
	Discharged battery	Check whether lights work. If battery is flat, remove and charge.
	Air leak at crankcase or worn crankshaft oil seals	Flood carburettors and check whether petrol is reaching the plugs.
Engine runs unevenly	Ignition and/or fuel system fault	Check as though engine will not start.
	Blowing cylinder head gasket	Oil leak should provide evidence. Replace gasket.
	Incorrect ignition timing	Check and if necessary adjust.
	Carburettors out of balance	Refer to Chapter 2 and adjust.
	Choked silencers	Remove baffles and clean.
Lack of power	Incorrect ignition timing	See above.
	Fault in fuel system	Check system and vent in filler cap.
	Choked silencers	See above.
White smoke from exhaust	Too much oil	Check oil pump setting.
	Engine needs rebore	Rebore and fit oversize pistons.
	Tank contains two-stroke petroil and not straight petrol	Drain and refill with straight petrol.
Engine overheats	Pre-ignition and/or weak mixture	Check carburettor settings, also grade of plugs fitted.
	Lubrication failure	Stop engine and check oil pump setting. Is oil tank dry?

53 Fault diagnosis: clutch

Symptom	Cause	Remedy
Engine speed increases but machine does not respond	Clutch slip	Check whether clutch adjustment still has free play. Check thickness of linings and renew if near wear limit.
Difficulty in engaging gears, gearchanges jerky and machine creeps forward, even when clutch is withdrawn fully	Clutch drag	Check clutch adjustment to eliminate excess play. Check whether clutch centre and outer drum have indented slots.
	Clutch assembly loose on mainshaft	Check tightness of retaining nut.
Operating action stiff	Bent pushrod	Renew.
	Dry pushrod	Lubricate.
	Damaged, trapped or frayed control cable	Check cable and renew if necessary. Make sure cable is lubricated and has no sharp bends.

54 Fault diagnosis: gearbox

Symptom	Cause	Remedy
Difficulty in engaging gears	Selector forks or rods bent	Renew.
	Broken springs in gear selector mechanism	Check and replace.
	Clutch drag	See previous Section.
Machine jumps out of gear	Worn dogs on ends of gear pinions	Strip gearbox and renew worn parts.
	Sticking camplate plunger	Remove plunger cap and free plunger assembly.
Kickstart does not return	Broken return spring	Remove right-hand crankcase cover and replace spring.
Kickstart slips or jams	Worn ratchet assembly	Remove right-hand crankcase cover, dismantle kickstart assembly and renew worn parts.
Gearchange lever does not return to normal position	Broken return spring	Remove right-hand crankcase cover and renew spring.

Chapter 2 Fuel system and lubrication

Contents

Specifications

Petrol tank
Capacity ... ...   3.6 Imp. gals (16.5 litre); [3.4 US gals (13.0 litre)]

Oil tank
Capacity ... ...   3.15 Imp pints (1.8 litre)

Carburettors
Make	Mikuni
Type	VM 28SC
Main jet	120 (115 USA)
Jet needle	5L1 - 3
Needle jet	P - 2
Throttle valve cutaway	2.5
Pilot jet	25
Air screw	1 - 1½
* Starter jet	70
Float valve seat	2.5
Float height	23 ± 2.5 mm

Left-hand carburettor only.

Oil pump
Minimum stroke tolerance	0.20 - 0.25 mm (0.008 - 0.010 in)
Oil viscosity	SAE 30

1 General description

The fuel system comprises a petrol tank from which petrol is fed by gravity, via a petrol tap with a built-in bowl-type filter, to the float chambers of the twin Mikuni carburettors.

For cold starting, the left-hand carburettor is fitted with a hand operated choke. This provides the rich mixture necessary for a cold start and can be opened as soon as the engine will accept full air under normal running conditions.

Unlike many two-strokes, the Yamaha RD400 twin does not require a petrol/oil mix for lubrication. Oil for lubricating the engine is contained within a separate, side-mounted oil tank, from which it is fed to an engine-driven oil pump, attached to the right-hand side of the crankcase. Oil from the pump is delivered to a drilling in the inlet passage of each cylinder barrel and is drawn into the engine with the incoming mixture. The

two-stroke engine depends on the compression of the incoming mixture before it is transferred to the combustion chamber via the transfer ports. Thorough lubrication of the bottom end of the engine as well as the cylinders and pistons is therefore essential. An added refinement is a direct link between the oil pump and the throttle so that the oil pump opening varies according to engine demand.

2 Petrol tank: removal and replacement

1 Although it is not necessary to remove the petrol tank when the engine unit is removed from the frame, better access is gained and there is less risk of damage to the painted surface if the tank is out of the way. Apart from occasions such as these, there is rarely any need to remove the tank unless rust has formed inside as the result of long storage or if it needs

repainting.

2   The tank is retained by rubber buffers at the front, which locate with cups attached either side of the frame top member, adjacent to the steering head lug. The rear of the tank is secured by two bolts passing through projections and supported on rubber buffers.

3   Before the petrol tank can be removed, it is necessary to detach the twin petrol pipes from the carburettor float chambers. Each is retained by a small wire clip. Detach the pipe that joins both halves of the petrol tank, on the underside. This too is retained at each end by a wire clip. The tank must be drained before it can be lifted from the frame, either by the petrol tap or by detaching one end of the underside cross pipe. The dual seat must be raised to gain access to the bolts at the rear of the tank.

4   When replacing the tank, check that the mounting rubbers are in good condition and are located correctly. Do not forget to replace and secure the pipe joining both halves of the petrol tank, before the tank is refilled.

3   Petrol tap: removal, dismantling and replacement

1   The petrol tap is secured to the left-hand underside of the petrol tank by two screws, which pass through the flange on the tap main body. The filter bowl threads into the main body of the petrol tap and can be removed by applying a spanner to the hexagon in the base of the bowl. There is a sealing washer between the filter bowl and the main body of the petrol tap to preserve a petrol-tight joint. No filter element is fitted within the bowl, this being placed above the tap and within the fuel tank itself. The filter bowl acts only as a sediment trap.

2   There is seldom need to disturb the main body of the petrol tap. In the event of leakage at the operating lever, the complete lever assembly can be dismantled (provided the petrol is first drained), with the main body undisturbed. Remove the single grub screw from beneath the lever housing boss and withdraw the lever. Note the 'O' ring and spring. The tap seal is of the plastic cone type and it is this component that must be renewed, if leakage occurs. The cone may be removed by inserting a pair of long nosed pliers into the tap bore. When refitting the lever assembly, note that the cone should be replaced with the two flow holes placed downwards and the lever fitted so that it points forwards, to the 'OFF' position.

3   The main body of the tap is secured to the underside of the petrol tank by two cross-head screws passing through a flange on the tap body. There is a sealing ring between the flange of the petrol tap body and the underside of the petrol tank, to preserve

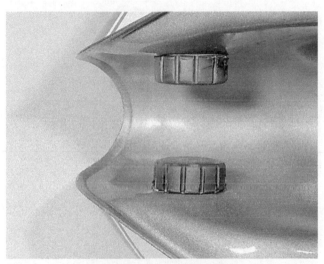

2.2 Fuel tank supported at front on rubber buffers

3.1a Sediment bowl fitted with 'O' ring

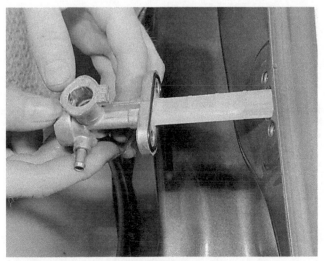

3.1b Main filter is fitted within petrol tank

3.2a Fuel tap lever is retained by a grub screw

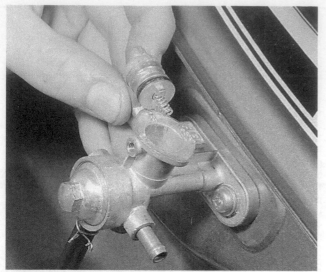

3.2b Note 'O' ring on lever and pressure spring for ...

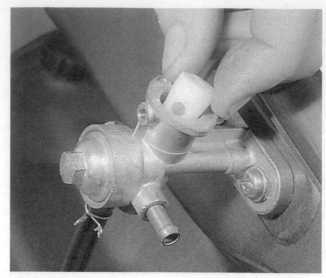

3.2c ... the nylon valve cone

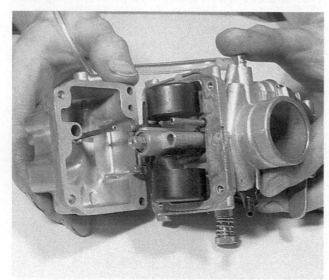

6.1 Remove the float chamber, held by four screws

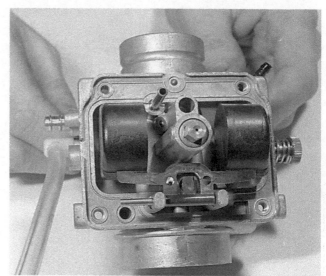

6.2 Displace the pivot pin to free float assembly

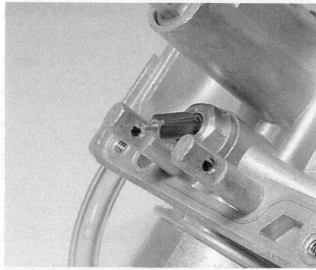

6.3a DO NOT mislay minute float needle

6.3b Needle seat is removable for cleaning

a petrol-tight joint. This gasket should be renewed if it is not in perfect condition, whenever the tap body is removed and replaced.

4   Before reassembling the petrol tap, check that all parts are clean, especially the two tubes (short tube reserve, long tube main feed) which extend into the petrol tank, the filter and filter bowl assembly. A new gasket should be fitted to the filter bowl assembly in order to effect a satisfactory seal.

5   Do not over-tighten any of the petrol tap components during reassembly. The castings are in a zinc-based alloy, which will fracture easily if over-stressed. Most leakages occur as the result of defective seals.

4   Petrol feed pipes: examination

1   The flexible pipes used for the various connections between the petrol tank and the carburettors are a thin-walled push-on type, secured by wire clips. Renewal is seldom required, unless it becomes hard or splits. Always replace the pipes in the same order because the ends sometimes take a permanent form of the fitting over which they are connected.

5   Carburettors: removal

1   Before removing the carburettors it is necessary to detach the petrol feed pipe at the point where it joins each float chamber. Pinch together the two 'ears' of the clip and slide the clip up the pipe so that the end of the pipe is free to be pulled from the float chamber connection.

2   Each carburettor is connected to the air cleaner by means of a short length of hose secured by a clip around each intake. Slacken the cross-head screw that retains each clip and pull the hoses away. The other end of each hose is a push fit over the inlet stub to the air cleaner housing, and is retained by screw clips in a similar way.

3   Unscrew the top of each mixing chamber and remove the top complete with the cable, throttle valve and needle assembly. It is a wise precaution to tape these parts to a near by frame tube so that they are not damaged as dismantling continues.

4   Slacken the cross-head screw in the clamp retaining each carburettor in its flexible rubber mounting stud. The carburettors can now be pulled away from their mountings and placed on the workbench for further dismantling. Note that they are inter-connected on the inside faces by a short rubber 'balance' pipe which interconnects the float chambers.

6   Carburettors: dismantling and examination

1   To separate the float chamber, invert the carburettor and remove the four small screws and spring washers holding the float chamber in position. There is a gasket between the float chamber body and the mixing chamber, to maintain a petrol-tight joint. The old gasket should be discarded.

2   The twin plastic floats will lift out of the float chamber as a single assembly, after pushing the pivot pin from position in the two posts.

3   Lift the float needle from its place in the needle seating. Put the needle to one side, in a safe place. It is very small and is easily lost. Persistent flooding is invariably caused by a leaking float, which will cause the petrol level to rise, by dirt on the float needle, or its seating. Renewal of the defective float is the only remedy in the case of a leaking float; it is not possible to effect a practical repair. Dirt on the float needle or its seating is best removed with a jet of compressed air.

4   Unscrew the main jet from the raised column in the base of the mixing chamber. The needle jet can then be pushed out towards the mixing chamber. This jet is subject to wear and should be renewed if the petrol consumption appears unduly high.

5   The throttle valve is still attached to the top of the mixing

6.4a Unscrew the main jet and ...

6.4b ... push the needle jet out towards the venturi side

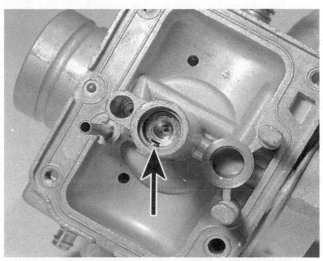

6.4c Note the peg with which the needle jet locates

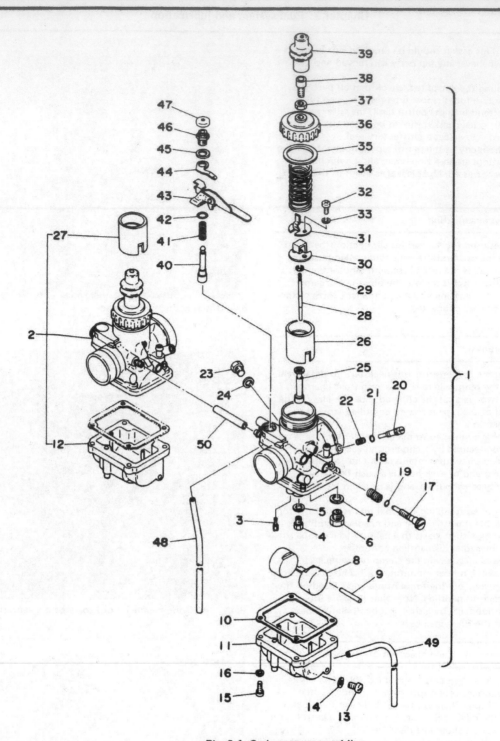

Fig. 2.1. Carburettor assemblies

1 LH carburettor	17 Throttle stop screw - 2 off	33 Spring washer - 2 off
2 RH carburettor	18 Spring - 2 off	34 Throttle return spring - 2 off
3 Pilot jet - 2 off	19 'O' ring - 2 off	35 Washer - 2 off
4 Main jet - 2 off	20 Pilot air screw - 2 off	36 Carburettor top - 2 off
5 Washer - 2 off	21 'O' ring - 2 off	37 Cable adjuster locknut - 2 off
6 Float valve assembly - 2 off	22 Spring - 2 off	38 Cable adjuster screw - 2 off
7 Washer - 2 off	23 Adjuster orifice plug - 2 off	39 Rubber boot - 2 off
8 Float assembly - 2 off	24 Washer - 2 off	40 Choke plunger
9 Float pivot pin - 2 off	25 Needle jet - 2 off	41 Return spring
10 Float bowl gasket - 2 off	26 LH throttle valve	42 'O' ring
11 LH float bowl	27 RH throttle valve	43 Choke lever
12 RH float bowl	28 Throttle needle - 2 off	44 Backing plate
13 Drain plug - 2 off	29 'E' clip - 2 off	45 Washer
14 Washer - 2 off	30 Needle plate - 2 off	46 Housing nut
15 Screw - 8 off	31 Spring seat - 2 off	47 Cap
16 Spring washer - 8 off	32 Screw - 4 off	48 Breather pipe - 2 off
		49 Breather pipe - 2 off
		50 Balance pipe

6.4d Pilot air screw is sealed by small 'O' ring

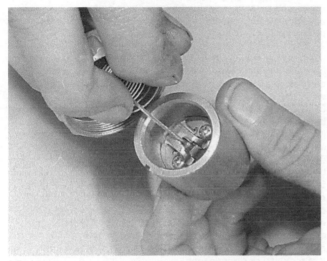

6.5a Lift throttle return spring to enable removal of ...

6.5b ... needle, needle holder and cable holder

chamber cap by means of the throttle cable and return spring. To release the throttle valve, lift the return spring and remove the metal seating that fits over the needle. This serves the dual function of providing a seating for the spring and by a bent tab which forms a locking device to prevent the throttle cable from being detached. When the seating is removed, the throttle cable nipple can be slipped out of the throttle valve and the throttle valve detached. The needle will lift out, with the retaining clip which holds it in the correct notch. When the needle jet is renewed, the needle should be renewed too, since they work in conjunction with one another.

6   Before reassembling the carburettor in the reverse order of that given for dismantling, make sure all the component parts are clean. Check that the needle is not bent, by rolling it on a sheet of plain glass. Examine the throttle slide; signs of wear will be evident on the polished outer surface.

7   Never use wire or any pointed instrument to clear a blocked jet or any of the internal air passages in the mixing chamber body. It is only too easy to enlarge the small precision drilled orifices and cause carburation changes which will prove very difficult to rectify. Always use compressed air; even a jet of air from a tyre pump should suffice.

8   When replacing the throttle valves, make sure the slot in the base of each valve registers with the projection inside the mixing chamber, so that the valve will seat correctly. It is also important to check that the needle suspended from each throttle valve has entered the needle valve, otherwise there is risk of damaging both the needle and the jet.

9   Similar advice to that given about the petrol tap applies to the carburettors, which are made of the same die-cast alloy. They will fracture or distort badly if force is used during reassembly.

7   Carburettors: checking the settings

1   The various jet sizes, throttle valve cutaway and needle position are predetermined by the manufacturer and should not require modification. Check with the Specifications list at the beginning of this Chapter if there is any doubt about the valves fitted.

2   Slow running is controlled by a combination of the throttle stop and pilot jet settings. Adjustment should be carried out as explained in the following Section. Remember that the characteristics of the two-stroke engine are such that it is extremely difficult to obtain a slow, reliable tick-over at low rpm. If desired, there is no objection to arranging the throttle stop so that the engine will shut off completely when the

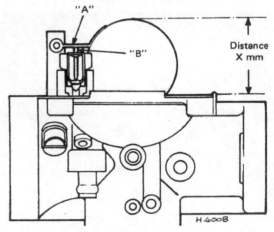

Fig. 2.2. Checking float level

A   Float tongue
B   Float valve
X   =   23 ± 2.5 mm (0.905 in)

throttle is closed. Unlike a petroil-lubricated engine, the oil used for engine lubrication is injected into the inlet passage of each cylinder barrel, behind the closed throttle slide. In consequence there is no risk of the engine 'drying up' when the machine is coasted down a long incline, if the throttle is closed.

3   As a rough guide, up to 1/8th throttle is controlled by the pilot jet, 1/8th to 1/4 by the throttle valve cutaway, 1/4 to 3/4 throttle by the needle position and from 3/4 to full by the size of the main jet. These are only approximate divisions, which are by no means clear cut. There is a certain amount of overlap between the various stages.

4   The normal setting of the pilot screw is approximately one and a half complete turns out from the closed position.
If the engine 'dies' at low speed, suspect a blocked pilot jet in the carburettor of the cylinder concerned.

5   Guard against the possibility of incorrect carburettor adjustments which will result in a weak mixture. Two-stroke engines are very susceptible to this type of fault, causing rapid overheating and often subsequent engine seizure. Changes in carburation leading to a weak mixture will occur if the air cleaner is removed or disconnected, or if the silencers are tampered with in any way. Above all, do not add oil to the petrol, in the mistaken belief that it will aid lubrication. The extra oil will only reduce the petrol content by the ratio of oil added, and therefore cause the engine to run with a permanently weakened mixture.

8   Synchronising twin carburettors

1   For optimum performance and even running, it is important that both carburettors are in phase, so that they open and close together and have the same settings throughout the entire throttle range. If this type of check is not carried out at regular intervals, one cylinder may do all the work, whilst 'carrying' the other. The usual symptoms of poor balance between the two carburettors are difficulty in starting and uneven running at low speeds, accompanied by a noticeable lag when accelerating.

2   To check whether the carburettors are correctly synchronised, remove the aligning orifice plug from the right-hand side of each carburettor and open the throttle fully. By viewing through the apertures, check that the lower edge of the aligning marks on the throttle valves align exactly with the lower edge of each aperture.

3   If alignment is incorrect adjust each carburettor independently by means of the cable adjuster in each carburettor

top. Finally, replace the orifice plugs.

4   Remove the plug cap from the left-hand spark plug and start the engine. Screw in (clockwise) the throttle stop screw of the right-hand carburettor until the engine will continue to run on one cylinder at a reasonably low speed. Check whether adjustment of the pilot screw from the recommended setting of one and a half complete turns out has any noticeable effect on even running and readjust the throttle stop screw as necessary. Stop the engine.

5   Repeat the above operation with the right-hand spark plug cap removed and adjust the left-hand carburettor. Stop the engine when the adjustment seems correct. Each cylinder should run independently at the same speed.

6   Re-start the engine with both plug caps connected. Most probably the tick-over will now be too fast, in which case the throttle stop screws of each carburettor can now be screwed out an equal amount (a little at a time) until the desired engine speed is attained. Normally this is in the region of 1,300 to 1,400 rpm, as indicated by the tachometer.

9   Air cleaner: removing, cleaning and replacing the element

1   The air cleaner is located beneath the nose of the dual seat, which must be raised to gain access to the element. The lid of the air cleaner box is retained by two wing bolts which must be unscrewed before the lid can be removed. The element (corrugated paper type), will lift out.

2   The element is cleaned by blowing it with compressed air or by lightly tapping it so that the loose dust on the surface will be displaced. Care is necessary when handling it because corrugated paper is impregnated with resin and is therefore easily damaged.

3   If the element is damp or contaminated in any way, it must be discarded and a new one fitted. This applies if it is perforated at any point. The filter is marked 'top' and 'front' and should be refitted accordingly.

4   Irrespective of its condition, the air cleaner element should be renewed every 3,000 miles.

5   Do not on any account run the machine with the element removed or with the air cleaner hoses disconnected. If this precaution is not observed, the engine will run with a permanently weak mixture, which will cause overheating and most probably engine seizure. The carburettors are jetted to compensate for the presence of the air cleaner element and the balance is upset when it is removed or disconnected.

8.2 Marks on both throttle valves must align simultaneously

8.6 Knurled throttle stop screw and slotted pilot air screw

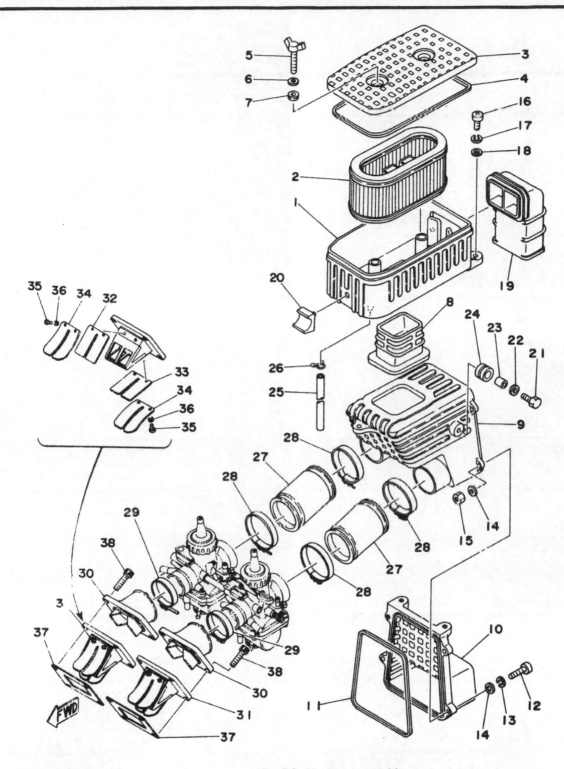

Fig. 2.3. Air cleaner assembly

1	Air cleaner element box	13	Spring washer - 4 off	26	Spring clip - 2 off
2	Air cleaner element	14	Plain washer - 8 off	27	Air hose - 2 off
3	Cover	15	Nut - 4 off	28	Hose clip - 4 off
4	Cover seal	16	Screw - 3 off	29	Screw clip - 2 off
5	Wing bolt - 2 off	17	Spring washer - 3 off	30	Inlet stub - 2 off
6	Plain washer - 2 off	18	Plain washer - 3 off	31	Reed valve block - 2 off
7	Insert - 2 off	19	Intake pipe	32	Top reed valve - 2 off
8	Transfer duct	20	Rubber buffer	33	Lower reed valve - 2 off
9	Silencing box	21	Bolt - 2 off	34	Reed valve stopper plate - 4 off
10	Silencing box cover	22	Plain washer - 2 off	35	Screw - 8 off
11	Cover seal	23	Spacer - 2 off	36	Spring washer - 8 off
12	Screw - 4 off	24	Grommet - 2 off	37	Valve block gasket - 2 off
		25	Drain hose - 2 off	38	Socket bolt - 8 off

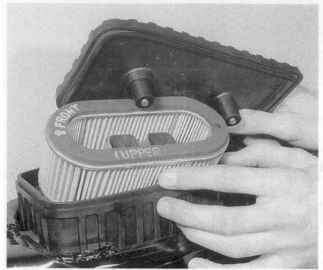

9.3 Air filter must be replaced in correct position

11.3a Inlet stub and reed valve is retained by four socket screws

11.3b Note heavy gasket between valve block and cylinder

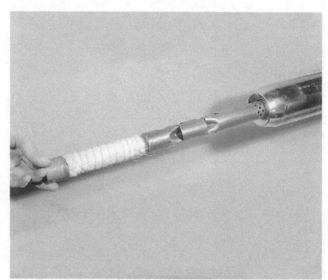

12.3a Baffles can be withdrawn for cleaning after ...

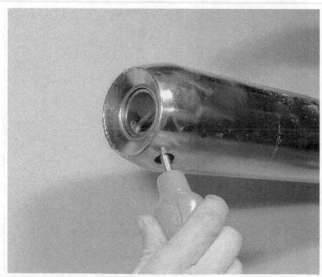

12.3b ... removal of screw at rear of silencer

12.6 The silencer/exhaust pipe connecting piece

10 Crankcase drain plugs

1 Unlike most two-stroke engines, the lower crankcase is not fitted with drain plugs with which to drain the contents of the two separate crankcase compartments.

2 If overflooding of the carburettors causes difficulty in starting the engine, the excess petrol vapour is best removed by unscrewing both spark plugs, turning off the petrol, and kicking over the engine with the throttle wide open until the engine has been thoroughly vented.

11 Reed valve induction system: mode of operation and examination

1 Of the various methods suitable for controlling the induction cycle of a two-stroke engine, Yamaha has adopted the reed valve, a non-mechanical device located between the carburettors and their respective cylinders. Each valve comprises two flexible stainless steel strips attached to a die-cast aluminium alloy casting. The strips seat on a gasket formed from heat and oil-resisting synthetic rubber, which is welded to the casting by the heat from the engine. A specially-shaped valve stopper, made of cold rolled stainless steel plate, acts as a form of stop, to control the extent to which the valves are free to move.

2 The valves open of their own accord as the piston commences to rise in the cylinder, creating a vacuum within the crankcase. Atmospheric pressure forces the valves open and causes a fresh fuel/air charge to be rammed into the crankcase. The existing mixture, already in the combustion is fully compressed and then ignited by the spark plug. The explosion drives the piston downwards again, expelling the exhaust gases through the exhaust port as it is uncovered by the falling piston. Although the reed valves had closed by the time the piston had reached the top of its stroke, they open again whilst the compressed charged in the crankcase passes into the combustion chamber via the transfer ports, through the inertia caused by the streams of fuel/air mixture entering the cylinder. This additional new charge is admitted by the seventh port and is not permitted to pass into the crankcase. Instead, it is used to expel the remainder of the exhaust gases out of the cylinder, so that they do not conflict with the incoming charge that is about to be compressed and fired. In other words this second action is of a scavenge nature only. From this point onwards the reed valves close until the piston is again on the ascent and the new charge in the combustion chamber is compressed and ignited. The reed valves open again to admit a new charge of fuel/air mixture to the crankcase and the whole cycle of operation is repeated.

3 The reed valves require no attention other than a wash with clean petrol during the course of an overhaul. They should be handled with great care and on no account dropped. If examination shows any signs of cracking or breakage, they should be renewed as a complete unit, without question. If any part of the valves should break off, it will be drawn into the engine where it is likely to cause serious damage.

4 Each cylinder has its own separate reed valve assembly.

12 Exhaust pipes and silencers: examining and cleaning the exhaust system

1 Two independent silencer/exhaust pipe units are fitted to the RD400 model, one for each cylinder. Each unit comprises a silencer and an exhaust pipe, interconnected by a flexible bonded rubber connector. The exhaust pipe is retained at the cylinder port by a flange secured on studs by two nuts. The silencer is supported by one bolt passing through the rear footrest mounting bracket and a second bolt passing into a threaded frame lug.

2 The parts most likely to require attention are the silencer baffles, which will block up with a sludge composed of carbon and oil if not cleaned out at regular intervals. A two-stroke

engine is very susceptible to this fault, which is caused by the oily nature of the exhaust gases. As the sludge builds up, back pressure will increase, with a resulting fall-off in performance.

3 There is no necessity to remove the exhaust system in order to gain access to the baffles. They are retained in the end of each silencer by a screw, reached through a hole cut in the underside of each silencer body, close to the end. When the screw is removed, the baffles can be withdrawn.

4 If the build-up of carbon and oil is not too great, a wash with a petrol/paraffin mix will probably suffice as the cleaning medium. Otherwise more drastic action will be necessary, such as the application of a blowlamp flame to burn away the accumulated deposits. Before the baffles are refitted, they must be thoroughly clean, with none of the holes obstructed.

5 When replacing the baffles, make sure the retaining screw is located correctly and tightened fully. If the screw falls out, the baffles will work loose, creating excessive exhaust noise accompanied by a marked fall-off in performance.

6 Do not run the machine without the baffles in the silencer or modify the baffles in any way. Although the changed exhaust note may give the illusion of greater power, the chances are that performance will fall off, accompanied by a noticeable lack of acceleration. There is also risk of prosecution by causing an excessive noise. The carburettors are jetted to take into account the fitting of silencers of a certain design and if this balance is disturbed, the carburation will suffer accordingly.

13 The lubrication system

1 Unlike many two-strokes, the Yamaha RD400 twin has an independent lubrication system for the engine and does not require the mixture of a measured quantity of oil to the petrol content of the fuel tank in order to utilise the so-called 'petroil' method. Oil of the correct viscosity (SAE 30) is contained in a separate oil tank mounted on the left-hand side of the machine and is fed to a mechanical oil pump on the right-hand side of the engine which is driven from the crankshaft by reduction gear. The pump delivers oil at a predetermined rate, via two flexible pipes, to oilways in the inlet passage of each cylinder barrel. In consequence, the oil is carried into the engine by the incoming charge of petrol vapour, when the inlet port opens.

2 The oil pump is also interconnected to the twist grip throttle, so that when the throttle is opened, the oil pump setting is increased a similar amount. This technique, pioneered by a British two-stroke manufacturer in the early 1930's, ensures that the lubrication requirements of the engine are always directly related to the degree of throttle opening. This facility is arranged by means of a control cable looped around a pulley in the end of the pump; the cable is joined to the throttle cable junction box at the point where the cable splits into two for the operation of each carburettor.

3 A dipstick is incorporated in the engine oil tank to aid oil level checking. In addition to this a sensor switch is also fitted, which illuminates a bulb in the warning lamp console when the oil falls below a pre-determined level.

14 Removing and replacing the oil pump

1 There is no necessity to remove the oil pump assembly unless the cover itself is damaged and has to be replaced. Under these circumstances the oil pump must be removed as a complete unit, so that it can be fitted to the new crankcase cover.

2 To remove the right-hand crankcase cover, follow the procedure given in Chapter 1, Section 5 (paragraphs 6 to 8) and Section 8 (paragraphs 1 and 2). There is no necessity to remove the engine from the frame in order to complete this operation, but the oil delivery pipes must be detached from the cylinder barrels.

3 The oil pump is secured to the cover by two cross-head screws, but before these can be removed, the drive pinion must first be detached. This is located on the inside of the outer cover

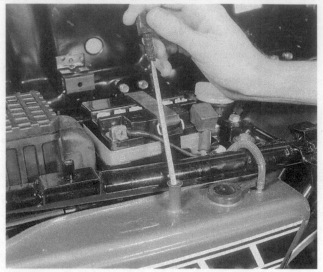

13.3 Dipstick enables easy checking of oil level in tank

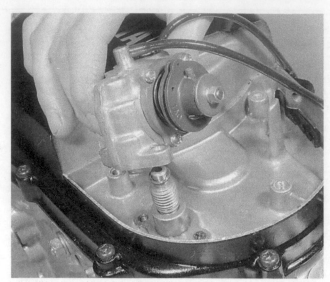

14.3 The oil pump is retained by two screws. Note shim on shaft

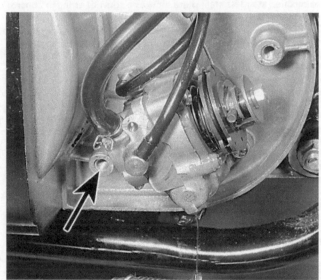

15.3 Bleed air from feed line after removing screw

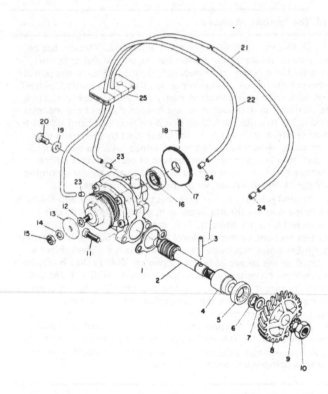

Fig. 2.4. Oil pump - component parts

1 Pump case gasket	14 Spring washer
2 Worm shaft	15 Nut
3 Dowel pin	16 Oil seal
4 Worm shaft outer bush	17 Starter plate
5 Oil seal	18 Split pin
6 Circlip	19 Bleed screw washer
7 Washer	20 Bleed screw
8 Pump drive gear pinion	21 Left-hand delivery pipe
9 Toothed washer	22 Right-hand delivery pipe
10 Nut	23 Delivery pipe clip - 2 off
11 Screw - 2 off	24 Delivery pipe clip - 2 off
12 Plunger shim	25 Oil pipe retainer
13 Adjusting plate	

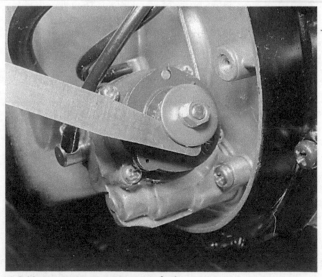

16.2 Measure pump stroke with feeler gauge

and is held by a nut and spring washer. When both are removed, the plastic pinion can be pulled off the oil pump driveshaft and the short metal rod passing through the shaft removed.

4 Unscrew the two cross-head screws from the other side of the crankcase cover, remove the circlip from the drive spindle, and lift away the oil pump, complete with the delivery pipes and the grommet through which they pass at the top of the crankcase cover.

5 Refit the oil pump to the replacement crankcase cover, using a new gasket at the oil pump/crankcase cover joint and a new oil seal behind the drive pinion. Replace and tighten the two cross-head mounting screws.

15 Bleeding the oil pump

1 It is necessary to bleed the oil pump every time the main feed pipe from the oil tank is removed and replaced. This is because air will be trapped in the oil line, no matter what care is taken when the pipe is removed.

2 Check that the oil pipe is connected correctly, with the retaining wire clip in position. Then remove the cross-head screw in the outer face of the pump body with the fibre washer beneath the head. This is the oil bleed screw. .

3 Check that the oil tank is not close to the refill level, then place a container below the oil bleed hole to collect the oil that is expelled as the pump is bled. Rotate the white plastic wheel at the base of the oil pump in a clockwise direction; this is the pinion with a milled edge with arrows showing the direction of rotation stamped on the face. Continue rotating the wheel until the oil expelled from the bleed hole is completely free from air bubbles, then replace the bleed screw and fibre washer. On some later models the plastic wheel is not fitted. Opening and closing the throttle will rotate the plunger pulley, which in turn will effect the bleeding operation. DO NOT replace the front portion of the crankcase cover until the pump setting has been checked, as described in the next Section.

16 Checking the oil pump setting

1 Make sure the twist grip is fully closed, then rotate the white plastic pinion used for bleeding the oil pump until the gap between the oil pump pulley at the opposite end of the casing and the body of the oil pump is at its maximum. It will be found that the pulley rises and falls as the pinion is turned, if light pressure is applied with the fingers to the end of the pulley. On the later machines where the knurled ring is not fitted, the pump should be rotated via the engine by means of the kickstart lever. Removal of the spark plugs will help to achieve smooth rotation.

2 Check the gap with a feeler gauge. It must be within the range 0.20 - 0.25 mm (0.008 in - 0.012 in) if the pump stroke is correct. To make adjustments, remove the locknut above the plate in the centre of the pulley and lift off the plate. If the clearance was too small, place the appropriate number of 0.1 mm (0.004 in) shims below the plate before replacing the plate and locknut. If the clearance is too great, remove the appropriate number shims from below the plate. Always re-check after replacing the plate and tightening down the locknut.

3 Adjust the throttle cable so that when the throttle is fully

16.4 With throttles fully open alignment on oil pump MUST be as shown

closed there is between 0.5 and 1 mm play. The adjuster is located close to the twist grip.

4 If the pump adjustment is now correct, the mark on the outer face of the oil pump pulley will be directly in line with the guide pin passing through the pulley boss, when the throttle is fully open.

5 Check that the pump pulley moves quite freely in each direction as the throttle is opened and closed. Replace the semi-circular end cover.

17 Removing and replacing the oil tank

1 The oil tank is secured to the frame by means of a flexible rubber mounting. The fixings are at the rear of the tank and take the form of a forward mounted bolt which threads into an insert in the tank body and a rear mounted screw that passes through the inner wall of the tank into a recessed area blanked off by the panel containing the capacity plate motif. The screw threads into an insert attached to the back of this panel and helps to hold it in position.

2 Before the tank can be removed, it is necessary to disconnect the breather tube attached to the vent immediately to the rear of the dipstick, the angled synthetic rubber connection between the filler cap and the inlet at the top inner face of the tank retained by a screw drive clip around the rubber hose, and the main feed pipe to the oil pump. When this latter pipe is disconnected by releasing and slipping the wire security clip down the hose and pulling off the connection, the tank can be drained of its oil content. Disconnect the leads from the oil level warning switch by separating the leads at the block connector.

3 It may be necessary to remove both the battery and the battery carrier to gain access to the screw drive clip around the filler hose, if the clip was positioned badly when the oil tank was fitted originally.

See next page for 'Fault diagnosis' - fuel system and lubrication system

18 Fault diagnosis: fuel system

Symptom	Cause	Remedy
Excessive fuel consumption	Air cleaner choked or restricted	Clean or renew element.
	Fuel leaking from carburettor	Check all unions and gaskets.
	Badly worn or distorted carburettors	Renew.
	Carburettor settings incorrect	Readjust. Check settings with Specifications.
Idling speed too high	Throttle stop screws in too far	Adjust screws.
	Carburettor tops loose	Tighten.
Engine sluggish. Does not respond to throttle	Back pressure in silencers	Check baffles and clean if necessary.
Engine dies after running for a short while	Blocked vent hole in filler cap	Clean.
	Dirt or water in carburettors	Remove and clean.
General lack of performance	Weak mixture; float needle sticking in seat	Remove float chambers and check needle seatings.
	Air leak at carburettors or leaking crankcase seals	Check for air leaks or worn seals.

19 Fault diagnosis: lubrication system

Symptom	Cause	Remedy
White smoke from exhausts	Too much oil	Check oil pump setting and reduce if necessary.
Engine runs hot and gets sluggish when warm	Too little oil	Check oil pump setting and increase if necessary.
Engine runs unevenly, not particularly responsive to throttle openings	Intermittent oil supply	Bleed oil pump to displace air in feed pipes.
Engine dries up and seizes	Complete lubrication failure	Check for blockages in feed pipes, also whether oil pump drive has sheared.

Note: *Lubrication failures will occur if a change is made from mineral oil to vegetable-based oils of the 'R' type (or vice-versa) if the engine is not stripped completely and all traces of the original oil removed. Mineral and vegetable oils do not mix, but under the action of heat form a rubber-like sludge that will quickly block the internal oilways.*

Chapter 3 Ignition system

Contents

Specifications

	UK	USA
Alternator		
Make	Hitachi	Mitsubishi
Model	LD 118 - 02	AZ 2015Y
Output	14v 12a @ 2000 rpm;	14v 15a @ 2000 rpm;
	18a @ 5000 rpm	20a @ 5000 rpm
Brushes	2	2
Brush length/wear limit	12 mm/7 mm	11 mm/6 mm
Brush spring tension	350 g	540 g
Stator coil resistance	0.54 ohms $\pm$ 10% at 20°C	0.46 ohms $\pm$ 10% at 20°C
Rotor coil resistance	4.53 ohms $\pm$ 10% at 20°C	5.5 ohms $\pm$ 15% at 20°C
Regulator unit		
Make	Hitachi	Mitsubishi
Model	TRIZ - 29	RFT 12M$_2$
Ignition coils		
Make	Hitachi	Hitachi
Model	CM11 - 53	CM11 - 53
Spark plug		
Make	NGK	NGK
Type	B - 8ES	B - 8ES
Reach	¾ in	¾ in
Gap	0.6 - 0.7 mm	0.6 - 0.7 mm
	(0.024 - 0.028 in)	(0.024 - 0.028 in)

1 General description

1 The spark necessary to ignite the petrol/air mixture in the combustion chambers is derived from the alternator attached to the left-hand end of the crankshaft. A twin contact breaker assembly, one set of points for each cylinder, determines the exact moment at which the spark will occur in the cylinder that is due to fire. As the points separate, the low tension circuit is interrupted and a high tension voltage is developed in the ignition coil, which passes across the points of the spark plug due to fire. This jumps the air gap and ignites the mixture under compression.

2 When the engine is running, the surplus current generated by the alternator is used to provide a regulated 12 volt supply for charging the battery, after it has been converted to direct current by the rectifier. If the battery is fully charged and the demand from the ignition and lighting circuits is low, this excess current is used to reduce the output from the alternator by the appropriate amount.

2 Alternator: checking the output

1 When the ignition switch is turned on, current from the battery flows through the closed points of the voltage regulator to the winding of the alternator rotor, via the positive brush which makes contact with the track on the face of the rotor. Immediately the rotor revolves, the magnetic field of the rotor windings induces current in the three separate coils of the stator assembly surrounding the rotor. This alternating current is used to operate the various electrical circuits and to charge the battery, after rectification to direct current.

2 It follows that good contact must be made between the carbon brushes and the tracks on the end of the alternator rotor with which they make contact. If the tracks, or slip rings as they are known, become oily or dirty, poor contact will result and the alternator output will drop.

3 The carbon brushes must be free in their holders. The brushes will be found in the stator cover that surrounds the alternator rotor, it also houses the twin contact breaker assembly. When

new, the brushes have a length of 12 mm (0.472 in). The brushes must be renewed when they wear to below 7 mm (0.275 in) in length. A wear mark is provided on each brush to aid measurement.

4   The output from the alternator can be checked only with specialised test equipment of the multi-meter type. It is unlikely that the average owner/rider will have access to this type of equipment or instruction in its use. In consequence, if the performance of the alternator is suspect, it should be checked by a Yamaha service agent or an auto-electrician.

3 Contact breakers: adjustment

1   To gain access to the contact breaker assembly remove the three cross-head screws securing the circular portion of the left-hand crankcase cover. Remove the cover. The twin contact breakers will be found on the stator cover enclosing the rotor of the crankshaft-driven alternator.

2   Rotate the engine until one set of points is fully-open. Examine the faces of the contacts. If they are dirty, pitted or burnt, it will be necessary to remove them for further attention, as described in Section 4 of this Chapter. Repeat this operation for the second set of contact points.

3   The correct contact breaker gap, when the points are fully open, is within the range 0.3-0.4 mm (0.012 in - 0.016 in). Adjustment is effected by slackening the screw holding the fixed contact breaker point in position and moving the point either closer or further away with a screwdriver inserted between the small upright post and the slot in the fixed contact plate. Make sure the points are fully open when this adjustment is made, or a false reading will result. When the gap is correct, tighten the retaining screw and re-check.

4   Repeat the procedure with the other set of contact breaker points.

4 Contact breaker points: removal, renovation and replacement

1   If the contact breaker points are burned, pitted or badly worn, they should be removed for dressing. If it is necessary to remove a substantial amount of material before the faces can be restored, new contacts should be fitted.

2   To remove the moving contact, slacken the screw and nut at the end of the return spring and remove the circlip from the post on which the contact pivots. The moving contact can now be lifted away complete with the return spring and the fibre heel bearing on the contact breaker cam.

2.3a Brush holder retained by single screw

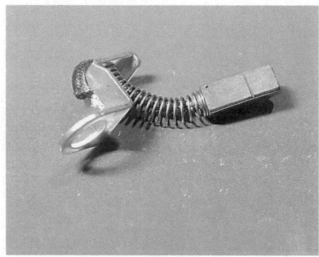

2.3b Wear indicator line on each brush

3.3 Check points gap using correct feeler gauge

4.2a Loosen nut to remove low tension lead before ...

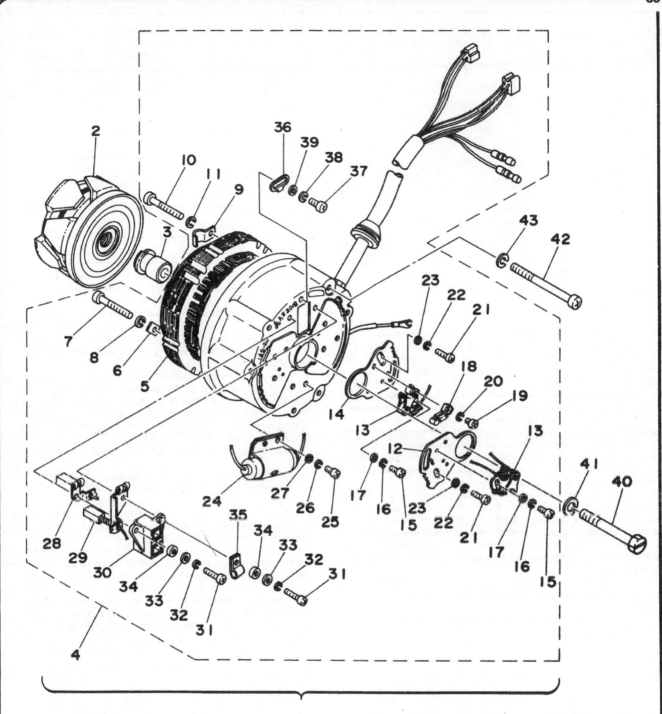

Fig. 3.1. Alternator and contact breaker assemblies

1   AC generator - complete
2   Rotor
3   Contact breaker cam
4   Stator assembly - complete
5   Stator coils
6   Plate washer - 2 off
7   Screw - 2 off
8   Spring washer - 2 off
9   Lead clamp
10 Screw
11 Spring washer
12 RH contact breaker base plate
13 Contact breaker assembly - 2 off
14 LH contact breaker base plate

15 Screw - 2 off
16 Spring washer - 2 off
17 Plain washer - 2 off
18 Lubricating wick
19 Screw
20 Spring washer
21 Screw - 4 off
22 Spring washer - 4 off
23 Plain washer - 4 off
24 Condenser
25 Screw - 2 off
26 Spring washer - 2 off
27 Plain washer - 2 off
28 Positive brush

29 Negative brush
30 Brush holder
31 Screw - 2 off
32 Spring washer - 2 off
33 Plain washer - 2 off
34 Spacer - 2 off
35 Lead clamp
36 Ignition timing plate
37 Screw
38 Spring washer
39 Plain washer
40 Cam centre bolt
41 Spring washer
42 Screw - 3 off
43 Spring washer - 3 off

4.2b ... removal of points set as a complete unit

3   To remove the fixed contact, remove the screw and nut that has already been slackened, so that the wires can be detached from the end of the mounting plate. Remove the screw that holds the mounting plate in position and lift away the plate complete with fixed contact.

4   When removing the wires from the fixed contact mounting plate, take particular note of the arrangement of the insulating washers. If they are replaced incorrectly, the points will be isolated electrically causing the ignition circuit to fail completely.

5   The points should be dressed with an oilstone or fine emery cloth. Keep them absolutely square during the dressing operation, otherwise they will make angular contact when they are replaced and will burn away rapidly as a result.

6   Replace the contacts by reversing the dismantling procedure, taking care to position the insulating washers in the correct sequence. Lightly grease the pivot post before replacing the moving contact and check that there is no oil or grease on the surface of the points. Place a few drops of oil on the lubricating wick that bears on the contact breaker cam, so that the surface is kept lubricated.

7   Re-adjust the contact breaker gap to the recommended setting, after verifying that the points are in their fully-open position.

8   Repeat the whole procedure for the other set of contact breaker points.

5 Condensers: location, removal and replacement

1   Condensers are included in the contact breaker circuit to prevent arcing across the contact breaker points as they separate. A condenser is connected in parallel with each set of contact points, and if a fault develops in either, or both condensers ignition failure is liable to occur.

2   If the engine is difficult to start, or if misfiring occurs, it is possible that a condenser is at fault. To check whether a condenser has failed, observe the points whilst the engine is running, after removing the circular portion of the left-hand crankcase cover. If excessive sparking occurs across one set of points and they have a blackened or burnt appearance, it may be assumed the condenser in that circuit is no longer serviceable.

3   The condensers are attached to the inside of the stator cover, which must be removed by unscrewing the three cross-head screws around the periphery holding the cover in position. Each is retained by a single screw through the strap soldered to the body of the condenser and by the lead wire attached to the screw and nut passing through the end of the moving contact return spring. Remove the screw and nut so that the terminal

end is freed. Because it is impracticable to repair a defective condenser, a new one must be fitted.

4   When the replacement is fitted, refit the stator cover and the circular portion of the left-hand crankcase cover. Note that it is extremely unlikely that both condensers will fail in unison; if total ignition failure occurs the source of the trouble should be sought elsewhere.

6 Condensers: testing

1   Without the appropriate test equipment, there is no alternative means of verifying whether a condenser is still serviceable.

2   Bearing in mind the low cost of a condenser, it is far more satisfactory to check whether it is malfunctioning by direct replacement.

7 Ignition coils: checking

1   Each cylinder has its own ignition circuit and if one cylinder misfires, one half of the complete ignition system can be eliminated immediately. The components most likely to fail in the circuit that is defective are the condenser and the ignition coil since contact breaker faults should be obvious on close examination. Replacement of the existing condenser will show whether the condenser is at fault, leaving by the process of elimination the ignition coil.

2   The ignition coil can best be checked using a multimeter set to the resistance position. Detach the orange lead and red/white lead at their snap connectors and detach the spark plug cap from the spark plug. Measure the primary winding resistance and the secondary winding resistance by connecting the multimeter as shown in the accompanying diagram.

The resistance values for each circuit should be as follows:

Primary coil resistance   1.4 ohms $\pm$ 10% at 20°C
Secondary coil resistance   6.6 ohms $\pm$ 20% at 20°C

Slight variation may be encountered if the ambient temperature departs greatly from that given. If the values differ from those given, the coil is faulty.

3   If a multimeter is not available, and by means of testing, the other components have been found to be satisfactory, the following method may be used to give an estimation of the coil's condition. Remove the suppressor cap and bare the inner wire. Remove the contact breaker cover and turn the engine over until the contact breaker points relevant to the coil to be tested are closed. Turn the ignition on and using an insulated screwdriver flick the points open and shut. If the bared end of the HT lead is held approximately 5 mm from an earthing point (the cylinder head) whilst this is done, a blue spark should jump the gap. If the spark is unable to jump a gap, or is yellowish in colour, the coil is probably at fault.

3   The ignition coils are sealed units and it is not possible to effect a satisfactory repair in the event of failure. A new coil must be fitted.

4   The ignition coils are mounted as a pair underneath the petrol tank. They bolt direct to a metal plate across the duplex top frame tubes and face in a rearward direction, parallel to the axis of the machine.

8 Ignition switch

1   The ignition switch is a multi-point switch which also controls the lighting circuits. It is bolted from the underside to the top yoke of the forks and is located immediately in front of the handlebars.

2   The switch is unlikely to malfunction during the normal service life of the machine and does not require any maintenance. If an ignition failure occurs and it would appear the switch may be responsible, the voltmeter test described in Section 7.2 will

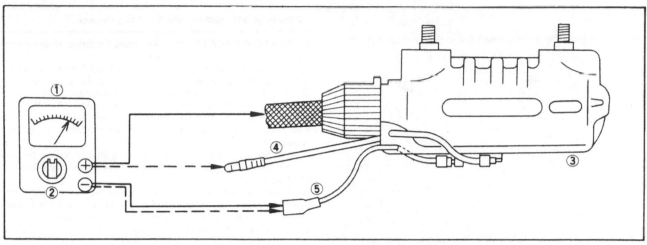

Fig. 3.2. Testing ignition coil continuity

1 Multi-meter   2 Set the tester on the 'Resistance' position   3 Ignition coil   4 Red/White   5 Orange

- - - - - → PRIMARY COIL RESISTANCE VALVE   ——→ SECONDARY COIL RESISTANCE VALVE

7.4 Ignition coils are located below petrol tank

9.1 Rough indication of timing given by index marks

confirm whether the switch is at fault. If the voltmeter does not give a reading when it is connected to the brown lead, with the ignition switched on and the contact breaker points closed, the switch is the source of the trouble, assuming the fuse in the electrical circuit has not blown. Replacement of the switch is the only remedy. Reconnection is easy, on account of the junction box connector used.

9 Ignition timing: checking and setting

1   If the ignition timing is correct, the contact breaker points of the cylinder about to fire must be on the verge of separation when the piston is 2.3 mm (0.090 in) before top dead centre. An approximate indication of the accuracy of the timing is given by the small pointer in one of the apertures of the stator cover of the alternator. This should line up with a scribe mark on the face of the rotor when the contact breaker points of the cylinder involved are on the point of separation. It must be stressed that this is only an approximate indication of the accuracy of the setting. Optimum performance depends on

timing the engine to a high degree of accuracy, to within ± 0.15 mm (0.0059 in) of the recommended setting on both cylinders.

2   To set the ignition timing with accuracy, remove the spark plug from the right-hand cylinder and fit a 14 mm dial gauge adaptor. Install the dial gauge and set it so that the dial shows a zero reading when the piston is **exactly** at top dead centre. Rotate the crankshaft backward (clockwise) check that the points for the right-hand cylinder (grey lead) are within the range 0.3-0.4 mm (0.012 in - 0.016 in) when they are fully open, then reverse the direction of rotation until the piston is 2.3 mm (0.090 in) **exactly** from top dead centre. If the timing is correct, the contact breaker points should be about to separate.

3   If the timing is not correct, adjust the contact breaker points by moving them as a unit either clockwise or anti-clockwise, depending on whether the opening point needs to be advanced or retarded. This is accomplished by slackening the two cross-head screws that pass through the elongated slots in the contact breaker base plate and moving the base plate with a screwdriver blade inserted between the short upright post and the notches in the edge of the base plate. When the adjustment is correct,

9.3 Timing adjustment screws

tighten both screws and re-check the setting.
4   Repeat this procedure for the left-hand cylinder, adjusting the points to which the orange lead is attached. This setting must be made with equal accuracy.
5   As a final check, attach the positive lead of a 0-20 volt range dc voltmeter to each moving contact in turn and earth the negative lead. Switch on the ignition and check that the voltmeter commences to show a reading when the piston is 2.3 mm (0.090 in) from top dead centre. Repeat for the other cylinder, with the other set of contact breaker points.

10 Spark plugs: checking and re-setting the gap

1   Two NGK B-8ES 14 mm spark plugs are fitted to the Yamaha RD400 twin as standard equipment.
2   The recommended spark plug gap is 0.6-0.7 mm (0.024 in - 0.028 in). Check the gap every 1,000 miles. To re-set, bend the outer electrode away from or closer to the centre electrode and check that a 0.025 in feeler gauge can be inserted. Never bend the central electrode, otherwise the insulator will crack, causing engine damage if the broken particles fall in whilst the engine is running.
3   After some experience the spark plug electrodes can be used as a reliable guide to engine operating conditions. See accompanying diagrams.
4   Always carry two spare spark plugs of the correct type. The plugs in a two-stroke engine lead a particularly hard life and are liable to fail more readily than when fitted to a four-stroke.
5   Never overtighten a spark plug, otherwise there is risk of stripping the threads from the cylinder head, especially as it is cast in light alloy. A stripped thread can be repaired without having to scrap the cylinder head by using a 'Helicoil' thread insert. This is a low-cost service, operated by a number of dealers.
6   Before replacing a spark plug into the cylinder head coat the threads sparingly with a graphited grease to aid future removal. Use the correct sized spanner when tightening plugs, otherwise the spanner may slip and damage the ceramic insulators. The plugs should be tightened sufficiently to seat firmly on their sealing washers, and no more.
7   Make sure that the plug insulating caps are a good fit and free from cracks. Apart from acting as an insulator from water and road dirt they contain the suppressor for eliminating radio and TV interference.

11 Fault diagnosis: ignition system

Symptom	Cause	Remedy
Engine will not start	No spark at plugs	Faulty ignition switch. Check whether current is reaching ignition coils.
	Weak spark at plugs	Dirty contact breaker points require cleaning. Contact breaker gaps have closed up. Re-set.
Engine starts, but runs erratically	Intermittent or weak spark on one cylinder	Locate defective cylinder and renew plug. If no improvement check whether points are arcing. If so renew condenser.
	Ignition over-advanced	Check ignition timing and if necessary, re-set.
	Plug lead insulation breaking down	Check for breaks in outer covering, especially near frame.
Engine difficult to start and runs sluggishly. Overheats	Ignition timing retarded	Check ignition timing and advance to correct setting.

Electrode gap check – use a wire type gauge for best results.

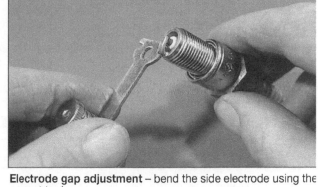

Electrode gap adjustment – bend the side electrode using the correct tool.

Normal condition – A brown, tan or grey firing end indicates that the engine is in good condition and that the plug type is correct.

Ash deposits – Light brown deposits encrusted on the electrode and insulator, leading to misfire and hesitation. Caused by excessive amounts of oil in the combustion chamber or poor quality fuel/oil.

Carbon fouling – Dry, black sooty deposits leading to misfire and weak spark. Caused by an over-rich fuel/air mixture, faulty choke operation or blocked air filter.

Oil fouling – Wet oily deposits leading to misfire and weak spark. Caused by oil leakage past piston rings or valve guides (4-stroke engine), or excess lubricant (2-stroke engine).

Overheating – A blistered white insulator and glazed electrodes. Caused by ignition system fault, incorrect fuel, or cooling system

Worn plug – Worn electrodes will cause poor starting in damp or cold conditions and will also waste fuel.

Chapter 4 Frame and forks

Contents

Specifications

Front forks

	UK	USA
Type	Oil damped telescopic	Oil damped telescopic
Damping oil capacity	145 cc (4.0 fl oz)	145 cc (4.0 fl oz)
Damping oil specification	SAE 10W/30 engine oil	SAE 10W/30 engine oil
Fork oil level from top of fork tube	389 ± 10 mm (15.3 in)	389 ± 10 mm (15.3 in)
Fork spring free length	349 mm (13.75 in)	422.5 mm (16.6 in)

Rear suspension

	UK	USA
Type	Swinging arm	Swinging arm

Suspension units

	UK	USA
Type	Oil damped, coil spring	Oil damped, coil spring
Spring free length	219 mm (8.6 in)	219 mm (8.6 in)
Swinging arm free play (max)	1.0 mm (0.0393 in)	1.0 mm (0.0393 in)

1 General description

The Yamaha 400 RD twin employs a frame and fork assembly of conventional design. The front forks are telescopic, with oil-filled, one-way damper units. The frame is of the full cradle type, employing duplex tubes. Rear suspension is provided by a swinging arm fork with replaceable bushes, controlled by hydraulically damped, adjustable, rear suspension units.

2 Front forks: removal from the frame

1 It is unlikely that the front forks will need to be removed from the frame as a complete unit unless the steering head bearings require attention or the forks are damaged in an accident.

2 Start by removing either the control cables from the handlebar control levers or the levers, complete with cables. The shape of the handlebars and the length of the control cables will probably dictate which method is used. Ensure that the four screws retaining the front brake master cylinder cover are tight before detaching the complete brake lever/master cylinder assembly from the handlebars. The lever assembly is retained by a clamp and two bolts. Tie the assembly to a suitable part of the frame that is not going to be disturbed so that the weight is not taken by the brake hose. It is quite possible to remove the forks without having to separate the brake components from one another. As this method eliminates the necessity of bleeding the front brake on reassembly, it is herein described.

3 Detach the handlebars from the top (crown) yoke. They are retained by two clamps, each of which is secured by two bolts. Unscrew the headlamp rim retaining screw which is located on the right-hand side of the shell and pull the complete rim/lens assembly away at the lower edge. Pull the wiring socket from the back of the reflector unit. Lift the dualseat and disconnect the negative lead (—) from the battery terminal followed by the positive (+) lead. This will isolate the electrical system and prevent short circuits occuring when the headlamp and instruments are removed. Detach the headlamp shell by removing the two bolts which pass through the shroud extensions (headlamp brackets) and screw into the shell.

4 If the machine is not already resting on the centre stand, support it in this manner on firm, level ground. Balance the machine so that the front wheel is clear of the ground and place some packing under the crankcase so that if the machine should inadvertently tip forward, it will not roll off the centre stand.

5 Remove the speedometer drive cable from the drive gearbox positioned to the left of the front wheel hub, on the wheel spindle.

The cable is retained by a knurled ring.

6   The front wheel can now be released by withdrawing the spindle, which passes through the left-hand fork leg and is retained by a castellated nut, and split pin. Note that it will be necessary to slacken the two nuts which secure the clamp around the head of the spindle, at the extreme end of the right-hand fork leg. The head of the spindle is drilled, so that a tommy bar can be inserted, to aid removal. Place one foot below the wheel to support the weight as the spindle is withdrawn, and lower the wheel so that it remains square, allowing the disc to leave the caliper unit easily.

7   Remove the single bolt and washer which retain the brake hose support clamp to the right-hand mudguard bracket. The front brake caliper can now be detached from the fork leg, to which it is secured by a domed nut and a dome headed bolt. Support the weight of the caliper unit as it is freed from the leg and then tie it to a suitable part of the frame so that the weight is not taken by the brake hose or brake pipe.

8   If desired, the front mudguard can be removed at this stage. It is secured to the inside of each fork leg by two bolts and washers which, when removed, will release the mudguard and stays as a complete unit. The speedometer cable will pass through the guide that keeps it clear from the front wheel.

9   Unscrew the large dome nut in the centre of the top yoke of the forks and remove it, together with the crown washer. Detach the drive cable from both the tachometer and the speedometer and the speedometer heads. Remove both instrument heads as a complete unit by detaching the mounting bracket from the top yoke. The bracket is retained by two integral studs secured at the underside by nuts. Note the rubber inserts, which help reduce vibration. Detach the wires from the warning lights at the snap connectors.

10   Slacken the pinch bolt at the rear of the top yoke and the pinch bolt holding the top of each fork leg. The yoke can now be lifted away, if necessary by lightly tapping the underside with a rawhide mallet to free it initially.

11   Whilst the forks are supported in this position it is a convenient opportunity to drain off the damping oil, especially if further dismantling is necessary after the forks have been removed from the frame. The drain screws (cross-head) are found at the lower end of each fork leg; place a container below each to catch the oil when the screw is removed and discard the oil.

12   Unscrew the slotted nut at the head of the steering column, using a 'C' spanner of the correct shape. As the nut is slackened, the forks will gradually ease away from the steering head,

2.5 Unscrew the speedometer cable retaining ring

2.6 Slacken the clamp nuts to withdraw spindle

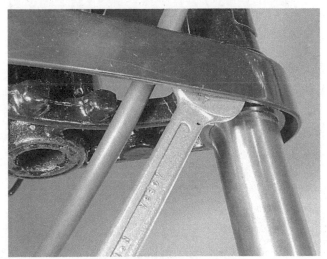

2.10 Loosen the fork leg pinch bolts

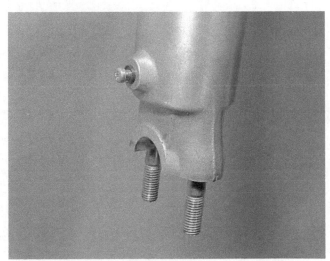

2.11 Drain plug fitted to each lower leg

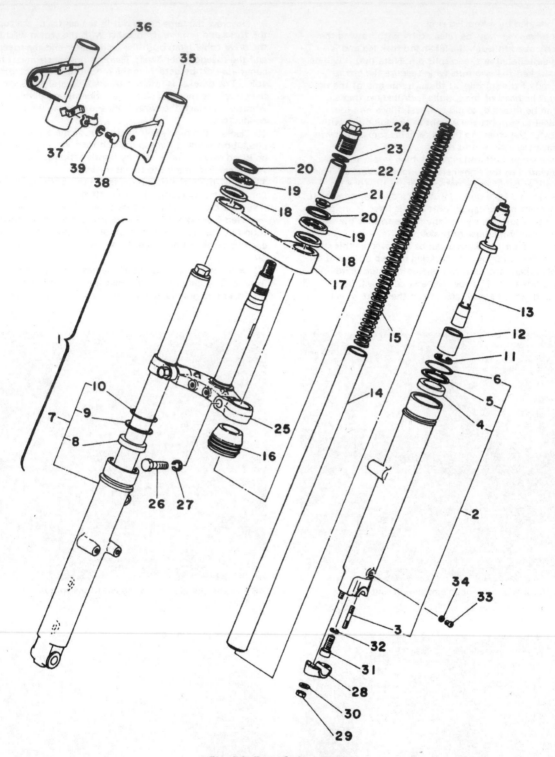

Fig. 4.1. Front fork assembly

1 Front fork assembly
2 LH lower leg - complete
3 Stud - 2 off
4 Oil seal
5 Backing washer
6 Circlip
7 RH lower leg - complete
8 Oil seal
9 Backing washer
10 Circlip
12 Damper piston - 2 off
13 Damper rod - 2 off

14 Fork stanchion (upper tube) - 2 off
15 Fork spring - 2 off
16 Dust excluder - 2 off
17 Lower yoke valance
18 Packing - 2 off
19 Shroud lower seat - 2 off
20 Shroud upper seat - 2 off
21 Spring upper seat - 2 off
22 Split spacer - 2 off
23 'O' ring - 2 off
24 Top bolt - 2 off
25 Lower yoke/steering stem assembly
26 Pinch bolt - 2 off

27 Spring washer - 2 off
28 Spindle clamp
29 Nut - 2 off
30 Plain washer - 2 off
31 Socket bolt - 2 off
32 Washer - 2 off
33 Drain plug - 2 off
34 Drain plug washer - 2 off
35 LH shroud
36 RH shroud
37 Cable clip - 2 off
38 Pinch bolt - 2 off
39 Spring washer - 2 off

uncovering the uncaged ball bearings of the steering head races. Make provision to catch the ball bearings as they are released; only the lower bearings will drop free since the upper bearings will most probably remain seated in the cup retaining them.

13 When the slotted nut has been removed from the steering column completely, the fork can be withdrawn from the lower end of the steering head as a complete unit. It may be necessary to raise the machine even higher during this operation, so that the fork stem will clear the steering head.

3 Front forks: dismantling

1 The fork legs can be dismantled individually, without need to disturb the steering head bearings. The preliminary dismantling is accomplished by following the procedure detailed in paragraphs 5 - 8 of the preceding Section, then continuing with the instructions given in this Section, after removing the chromium plated bolts from the top of each fork leg and draining off the damping oil.

2 If both fork legs are to be dismantled, strip them separately, using an identical procedure. There is less chance of unwittingly interchanging parts if this approach is adopted,

3 Slacken the pinch bolt through the lower fork yoke and pull the complete fork leg from the assembly, leaving the upper fork shroud in position. Invert the fork with the spring still in position and using an hexagonal allen key, remove the socket screw recessed into the curved portion of the lower fork end, through which the wheel spindle normally passes. The spring pressure is essential, to prevent the fork damper unit from rotating whilst this socket screw is removed. If, in spite of the spring pressure, the damper rod still rotates, it will be necessary to manufacture a special locking tool to hold the rod steady. The tool may be fabricated from a length of mild steel rod or square section bar, the end of which should be ground to a screwdriver type profile. The tip may be engaged with one of the two flats milled on the inner head of the damper rod. See the accompanying photograph for an example of a fabricated tool.

4 Remove the top bush and spring from within the fork leg, then re-invert the fork leg and prise off the dust cover, where the sliding action of the fork occurs. Remove the dust seal and the circlip within the top end of the lower fork leg retaining the oil seal and oil seal washer. The fork tube can now be pulled away from the lower fork leg and separated.

5 To remove the damper unit, invert the fork leg and detach

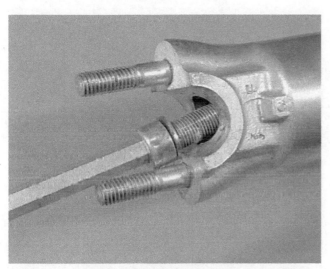

3.3a Unscrew the socket bolt which retains the damper rod

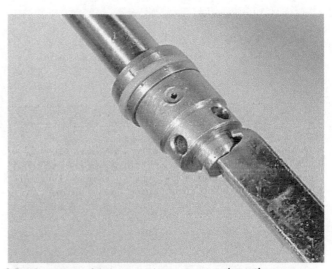

3.3b If necessary fabricate tool to prevent rod rotation

3.3c Remove the chrome top bolt and ...

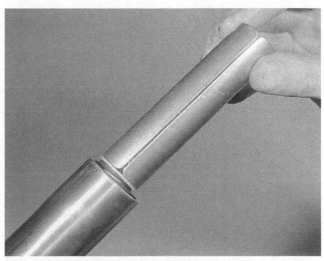

3.4a ... withdraw the spring spacer and ...

3.4b ... the fork spring itself

3.4c Separate the upper tube from the lower leg

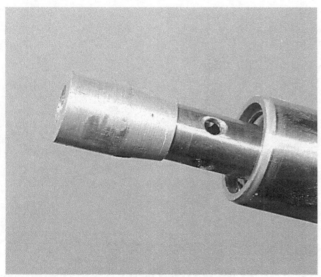

3.5a Pull the damper seat from position and ...

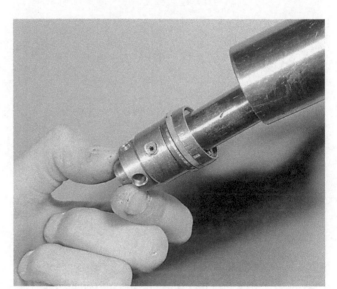

3.5b ... invert the tube to allow rod removal

3.5c Remove the circlip in the fork upper tube to ...

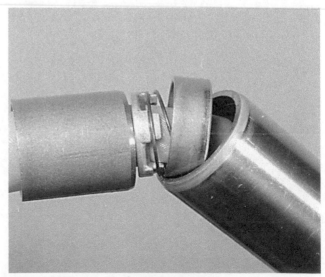

3.5d ... permit removal of valve components

3.5e Oil seal is retained by 'C' clip and backing washer

the circlip inside the tapered bottom end. The damper assembly, complete with piston, can now be drawn out as a complete assembled unit. No further dismantling is possible.

4 Steering head bearings: examination and renovation

1 Before reassembly of the forks is commenced, examine the steering head races. The ball bearing tracks of the respective cup and cone bearings should be polished and free from indentations or cracks. If wear or damage is evident, the cups and cones must be renewed as a complete set. They are a tight press fit and should be drifted out of position.
2 Ball bearings are cheap. If the originals are marked or discoloured, they should be renewed. To hold the steel balls in position during reassembly, pack the bearings with grease. Note that each race contains only nineteen ¼ in ball bearings. There is space for the addition of one extra ball, but this must be left empty to prevent the ball bearings from skidding on one another, a situation which would greatly accelerate the rate of wear.

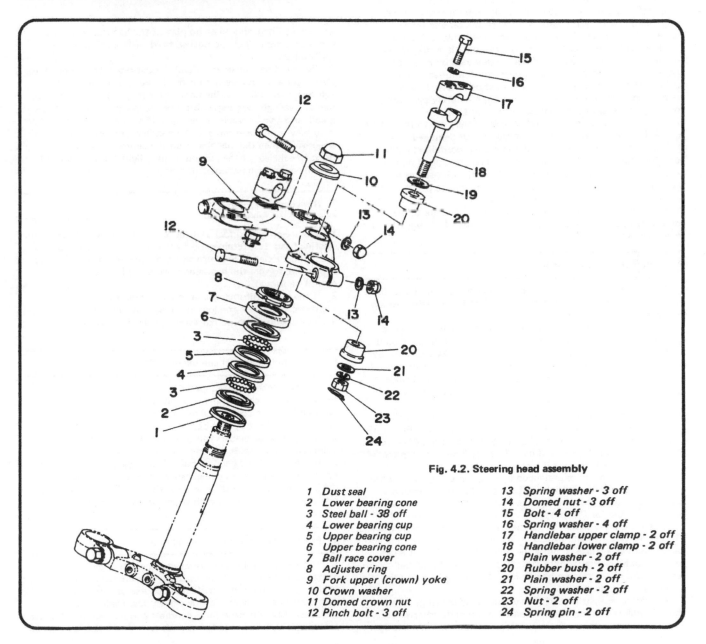

Fig. 4.2. Steering head assembly

1 Dust seal
2 Lower bearing cone
3 Steel ball - 38 off
4 Lower bearing cup
5 Upper bearing cup
6 Upper bearing cone
7 Ball race cover
8 Adjuster ring
9 Fork upper (crown) yoke
10 Crown washer
11 Domed crown nut
12 Pinch bolt - 3 off
13 Spring washer - 3 off
14 Domed nut - 3 off
15 Bolt - 4 off
16 Spring washer - 4 off
17 Handlebar upper clamp - 2 off
18 Handlebar lower clamp - 2 off
19 Plain washer - 2 off
20 Rubber bush - 2 off
21 Plain washer - 2 off
22 Spring washer - 2 off
23 Nut - 2 off
24 Spring pin - 2 off

5 Front forks: examination and renovation

1   The parts most liable to wear over an extended period of service are the wearing surfaces of the fork stanchion and lower leg, the damper assembly within the fork tube and the oil seal at the sliding joint. Wear is normally accompanied by a tendency for the forks to judder when the front brake is applied and it should be possible to detect the increased amount of play by pulling and pushing on the handlebars when the front brake is applied fully. This type of wear should not be confused with slack steering head bearings, which can give identical results.

2   Renewal of the worn parts is quite straightforward. Particular care is necessary when renewing the oil seal, to ensure that the feather edge seal is not damaged during reassembly. Both the seal and the fork tube should be greased, to lessen the risk of damage.

3   After an extended period of service, the fork springs may take a permanent set. The free length is 349 mm (13.75 in) or 422.5 mm (16.6 in) in the case of US specification models. It is wise to fit new components if the overall length has decreased. Always fit new springs as a pair, NEVER separately.

4   Check the outer surface of the fork tube for scratches or roughness. It is only too easy to damage the oil seal during reassembly, if these high spots are not eased down. The fork tubes are unlikely to bend unless the machine is damaged in an accident. Any significant bend will be detected by eye, but if there is any doubt about straightness, roll the tubes on a flat surface. If the tubes are bent, they must be renewed. Unless specialised repair equipment is available, it is rarely practicable to straighten them to the necessary standard.

5   The dust seals must be in good order if they are to fulfil their proper function. Replace any that are split or damaged.

6   Damping is effected by the damper units contained within each fork tube. The damping action can be controlled within certain limits by changing the viscosity of the oil used as the damping medium, although a change is unlikely to prove necessary except in extremes of climate.

7   Note that the forks are not fitted with renewable bushes. If wear develops, the stanchions and/or the lower fork legs will have to be renewed.

6 Front forks: replacement

1   Replace the front forks by reversing either of the dismantling procedures described in Sections 2 and 3 of this Chapter, which-ever the more appropriate.

2   Before fully tightening the front wheel spindle, right-hand spindle clamp, fork yoke pinch bolts and the chromium plated bolts in the top of each fork leg, bounce the forks several times to ensure that they work freely and settle down in their original positions. Complete the final tightening from the front wheel spindle upward, and do not forget to fit and open the split pin through the wheel spindle nut.

3   Do not forget to add the correct amount of damping oil to each fork leg before the bolts in the top are tightened. Each fork should be filled with 145 cc SAE 10W/30 engine oil. Check that the drain plugs at the front of each fork leg have been replaced and tightened, before the oil is added!

4   Difficulty is often experienced when attempting to draw the fork inner tube into position in the top yoke during assembly, even though the tube does not have a taper fit. A Yamaha service tool is available for this purpose, in the form of a portion of threaded rod on which a 'T' handle has been brazed across the top. The rod screws into the thread on the inside of the fork tube and can be used as a guide to draw the tube into position. It is easy to construct a similar tool or if time is short, to thread the tapered end of the broom handle into the fork tube as a temporary expedient.

5   Check the adjustment of the steering head bearings before the machine is used on the road and again shortly afterwards, when they settle down. If the bearings are too slack, fork judder

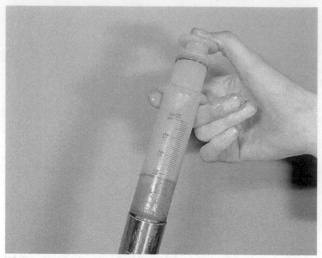

6.3 Do not forget to replenish forks with damping fluid

will occur. There should be no play at the headraces when the handlebars are pulled and pushed hard, with the front brake applied hard.

6   Overtight headraces are equally undesirable. It is possible to place a pressure of several tons on the head bearings by over-tightening, even though the handlebars may seem to turn quite freely. Overtight bearings will cause the machine to roll at low speeds and give imprecise steering. Adjustment is correct if there is no play in the bearings and the handlebars swing to full lock either side when the machine is on the centre stand with the front wheel clear of the ground. Only a light tap on each end should cause the handlebars to swing.

7 Steering head lock

1   The steering head lock is attached to the left-hand side of the steering head. It is retained by a rivet. When in a locked position, the plunger extends and engages with a portion of the steering head stem, so that the handlebars are locked in position and cannot be turned.

2   If the lock malfunctions, it must be renewed. A repair is impracticable. When the lock is changed it follows that the key must be changed too, to correspond with the new lock.

8 Handlebars: removal and examination

1   The handlebars are retained on the fork upper yoke by two supporting split clamps, the lower half of each passing through the yoke and being secured by a nut and split pin. The upper clamp halves are retained by two bolts each.

2   Before the handlebars can be removed, the controls must be detached or the cables disconnected at the levers.

3   If the handlebars become bent in an accident, they should be renewed rather than straightened. Bars are notoriously difficult to re-bend to their original shape and in any case the chrome plate will have stretched and will soon start flaking.

9 Frame: examination and renovation

1   The frame is unlikely to require attention unless it is damaged as the result of an accident. In many cases, replacement of the frame is the only satisfactory course of action, if it is badly out of alignment. Comparatively few frame repair specialists have the necessary mandrels and jigs essential for the accurate re-setting of the frame and, even then there is no means

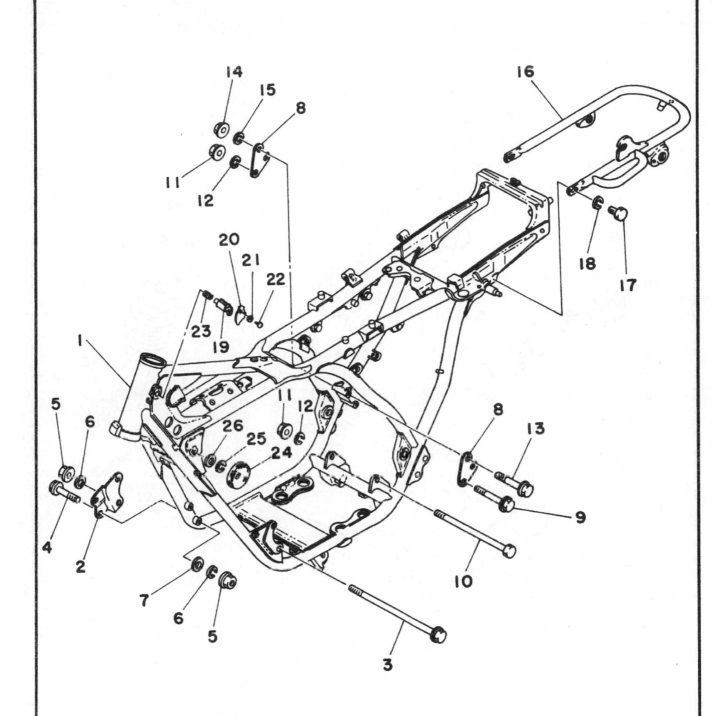

Fig. 4.3. Frame assembly

1	Frame unit	9	Bolt	18	Spring washer - 2 off
2	Engine front mounting plate	10	Bolt	19	Steering lock
3	Bolt - 2 off	11	Nut - 2 off	20	Lock cover
4	Bolt - 2 off	12	Spring washer - 2 off	21	Washer
5	Nut - 4 off	13	Bolt - 2 off	22	Rivet
6	Spring washer - 4 off	14	Nut - 2 off	23	Conical spring
7	Plain washer - 2 off	15	Spring washer - 2 off	24	Reflector lens - 2 off
8	Engine rear mounting plate - 2 off	16	Rear stay	25	Spring washer - 2 off
		17	Bolt - 2 off	26	Plain washer - 2 off

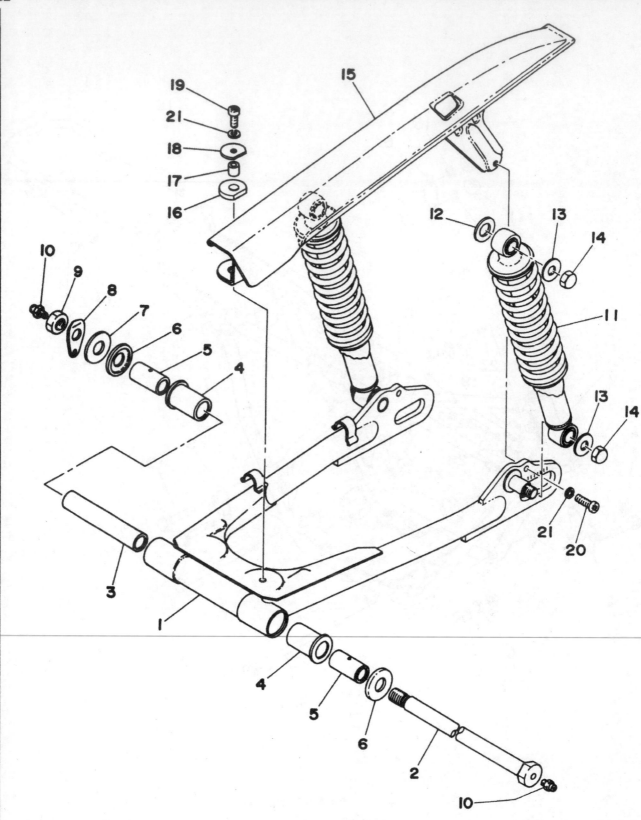

Fig. 4.4. Rear swinging arm assembly

1	Swinging arm fork	8	Tab washer	15	Chain case
2	Swinging arm pivot shaft	9	Nut	16	Damper rubber
3	Bush spacer	10	Grease nipple - 2 off	17	Spacer
4	Outer bush - 2 off	11	Suspension unit - 2 off	18	Washer
5	Inner bush - 2 off	12	Plain washer - 2 off	19	Screw
6	Thrust cover - 2 off	13	Plain washer - 4 off	20	Screw
7	Shim - A/R	14	Cap nut - 4 off	21	Spring washer - 2 off

of assessing to what extent the frame may have been overstressed such that a later fatigue failure may occur.

2 After a machine has covered an extensive mileage, it is advisable to keep a close watch for signs of cracking or splitting at any of the welded joints. Rust can cause weakness at these joints particularly if they are unpainted. Minor repairs can be effected by welding or brazing, depending on the extent of the damage found.

3 A frame out of alignment, will cause handling problems and may even promote 'speed wobbles' in a particular speed range. If misalignment is suspected as the result of an accident, it will be necessary to strip the machine so that the frame can be checked, and if needs be, renewed.

10 Swinging arm rear fork: dismantling, examination and renovation

1 The rear fork of the frame assembly pivots on a detachable bush within each end of the fork cross member and a pivot shaft which passes through frame lugs and the centre of each of the two bushes. It is quite easy to renovate the swinging arm pivots when wear necessitates attention.

2 To remove the swinging arm fork, first place the machine on the centre stand, then detach the final drive chain, preferably whilst the spring link is resting in the teeth of the rear wheel sprocket. It is advisable to remove the chainguard, which is attached at the rear by a cross-head screw that threads into the left-hand chainstay, immediately to the rear of the lower end of the suspension unit mounting. The forward end of the chain-guard is attached to the left-hand side of the swinging arm fork by a cross-head screw which passes through a rubber mounting.

3 Detach the rear brake torque arm at the brake plate by removing the split pin, then unscrewing and detaching the nut and washer. Remove the split pin from the left-hand end of the rear wheel spindle, then the castellated nut. Support the weight of the wheel with one foot and withdraw the wheel spindle. Lower the wheel away from the fork, preparing to catch either spindle spacer if it falls free. The chain adjuster on the left-hand fork end will probably fall from position.

4 The rear brake caliper unit is mounted on a cast plate, which is attached to the right-hand fork end by a hollow bolt through which the wheel spindle passes. The caliper/mounting plate assembly can be removed as a single unit and suspended from a suitable portion of frame tube - where it will be out of harms way - without need for disconnecting the brake hose. Support the weight of the caliper while undoing the retaining hollow bolt and nut.

5 Remove both rear suspension units. Each is retained in position by a domed nut and washer. When the nuts and washers have been removed, the suspension units can be pulled off their mounting studs. Knock down the ear of the tab washer which secures the nut on the right-hand end of the swinging arm pivot bolt. Loosen and remove the nut. Pull the pivot shaft away from the left-hand side of the machine, then pull the swinging arm fork from the rear. It will come away complete with the thrust covers and any shims that have been added to take up play. Note the position of the shims and their thickness so that they are replaced in the correct order.

6 Remove both thrust covers and withdraw the inner bushes from each end of the cross member. Note that there is a much longer distance piece between them which serves no load bearing function.

7 Wear will take place in both the bushes and the pivot shaft, which should be renewed together, never separately. It is also advisable to renew the shaft if it is out of true, irrespective of the condition of the bushes. The outer bushes may be drifted from position using a suitable long shanked drift passed through the cross member.

8 Reassemble the swinging arm fork by reversing the dis-mantling procedure. Grease the pivot shaft and the bearings liberally, prior to assembly and check that the grease nipples in both ends of the pivot shaft are unobstructed. When reassembly

10.5a Knock back tab washer and loosen pivot nut

10.5b Carefully drift pivot shaft from position

10.5c Complete swinging arm can be tilted and lifted away

10.6 Inner bushes are a push fit. Note side shim

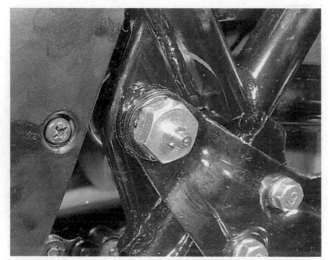

10.8 Grease pivot shaft thoroughly, after assembly

14.1 Front footrests are mounted as a unit on rubber dampers

is complete, apply a grease gun to both ends of the pivot shaft and continue pumping until grease can be seen emerging from the ends of the pivot joint.

9   Apart from causing a machine to fail the DOE test, worn swinging arm pivot bearings will give imprecise handling, with a tendency for the rear of the machine to twitch or hop. The play can be detected by placing the machine on the centre stand with the rear wheel clear of the ground, by pulling and pushing alternately on the fork ends.

11 Rear suspension units: examination

1   Rear suspension units of the hydraulically damped type are fitted to the RD 400 Yamaha twin. They can be adjusted to give five different spring loadings, without removal from the machine.
2   Each rear suspension unit has two peg holes immediately above the adjusting notches, to facilitate adjustment. Either a 'C' spanner or the screwdriver supplied with the original tool kit can be used to turn the adjusters. Turn clockwise to increase the spring tension and stiffen up the rear suspension.
    The recommended settings are:-
    Position 1 (least tension) for normal solo riding and
    position 5 (greatest tension) for high speed riding or when carrying a heavy load. The intermediate settings may be used for varying conditions, as required.
3   The suspension units are sealed and there is no means of topping up or changing the damping fluid. If the damping fails or if the unit leaks, renewal is necessary.
4   In the interests of good roadholding it is essential that both suspension units have the same load setting. If a renewal is necessary, the units must be replaced as a matched pair.

12 Centre stand: examination

1   The centre stand is attached to lugs welded to the bottom of the rearmost cross member of the duplex tube frame, below the engine. The pivot is bushed and the stand is retained by a pivot shaft with a split pin through one end. An extension spring, fitted on the right-hand side, keeps the stand in the fully-retracted position when the machine is in use.
2   Check that the return spring is in good condition and correctly located. If the stand drops when the machine is in motion it may catch in some obstacle and unseat the rider.

13 Prop stand: examination

1   A prop stand is fitted, for occasional parking when it is not desired to use the centre stand. The prop stand pivots from a lug welded to the lower left-hand tube, close to the crankcase of the engine. A bolt and nut pass through both the prop stand and the lug to act as the pivot; the nut is retained by a split pin that passes through a drilling in the end of the bolt. An extension spring returns the stand to the retracted position, immediately the weight is taken from the prop stand.
2   Check that the split pin has not been omitted from the pivot bolt and that the nut is tight. Check also that the extension spring is not overstretched or worn at the end connection. An accident is almost inevitable if the prop stand should fall whilst the machine is on the move.

14 Footrests: examination and renovation

1   The footrests are made as a complete unit, which is held to lugs welded to the underside of the duplex frame tubes below the engine by four nuts and bolts. The nuts are retained by split pins that pass through drillings in the ends of the bolts; rubber anti-vibration mountings give the footrests a certain amount of flexibility.
2   Both pairs of footrests are pivoted on clevis pins and spring

loaded in the down position. If an accident occurs, it is probable that the footrest peg will move against the spring loading and remain undamaged. A bent peg may be detached from the mounting, after removing the clevis pin securing split pin and the clevis pin itself. The damaged peg can be straightened in a vice, using a blowlamp flame to apply heat at the area where the bend occurs. The footrest rubber will, of course, have to be removed as the heat will render it unfit for service.

15 Rear brake pedal: examination and renovation

1   The rear brake pedal pivots through the right-hand silencer stay, to which a short length of tube is welded. The shaft carrying the brake arm is splined, to engage with splines of the rear brake pedal. The pedal is retained to the shaft by a simple pinch bolt arrangement.
2   If the brake pedal is bent or twisted in an accident, it should be removed by slackening the pinch bolt and straightened in a manner similar to that recommended for the footrests in the preceding Section.
3   Make sure the pinch bolt is tight. If the lever is a slack fit on the splines, they will wear rapidly and it will be difficult to keep the lever in position.

16 Kickstart lever: examination and renovation

1   The kickstart lever is splined and is secured to its shaft by means of a pinch bolt. The kickstart crank swivels so that it can be tucked out of the way when the engine is started. It is held in position on the swivel by a washer and circlip. A spring loaded ball bearing locates the kickstart arm in either the operating or folded position; if the action becomes sloppy it is probable that the spring behind the ball bearing needs renewing. It is advisable to remove the circlip and washer occasionally, so that the kickstart crank can be detached and the swivel greased.
2   It is unlikely that the kickstart crank will bend in an accident unless the machine is ridden with the kickstart in the operating and not folded position. It should be removed and straightened, using the same technique as that recommended for the footrests in Section 14.2.

17 Dualseat: removal and replacement

1   The dualseat is attached to the right-hand frame tube by means of two pivots on which it hinges. A catch on the left-hand frame tube locks the dualseat in position, under normal riding conditions.
2   To release the dualseat from the machine, lift the catch and lift the dualseat so that the pivots on the right-hand side are exposed. If the split pin though the pivots are removed and the pivot pins withdrawn, the dualseat can be lifted away.

18 Speedometer and tachometer heads: removal and replacement

1   The tachometer and speedometer heads are rubber mounted on a common support bracket, which is retained on the fork top (crown) yoke by two integral studs secured on the underside by nuts.
2   The instrument heads may be detached from the mounting plate individually, after the cable has been detached by unscrewing the knurled rings and the two dome nuts and washers removed from the underside.
3   Each instrument head rests in a shell where it is supported by a rubber cushioning ring and secured by two dome nuts. Remove the nuts and washers and separate the instrument from the shell. It will be necessary to remove the bulbs

from the case of each instrument head by pulling the bulb-holders from their seatings; each is retained by a rubber cup.
4   Do not misplace the rubber cushion interposed between the mounting bracket and the instrument case to damp out the undesirable effects of vibration.
5   Apart from defects in either the drive or the drive cable, a speedometer or tachometer that malfunctions is difficult to repair. Fit a new one, or alternatively entrust the repair to a component instrument repair specialist.
6   Remember that a speedometer in correct working order is a statutory requirement in the UK and many other countries. Apart from this legal requirement, reference to the odometer reading is the best means of keeping in pace with the maintenance schedule.

19 Speedometer and tachometer drive cables: examination and maintenance

1   It is advisable to detach both cables from time to time in order to check whether they are lubricated adequately, and whether the outer coverings are compressed or damaged at any point along their run. Jerky or sluggish movements can often be attributed to a cable fault.
2   For greasing, withdraw the inner cable. After wiping off the old grease, clean with a petrol-soaked rag and examine the cable for broken strands or other damage.
3   Regrease the cable with high melting point grease, taking care not to grease the last six inches at the point where the cable enters the instrument head. If this precaution is not observed, grease will work into the head and immobilise the movement.
4   If either instrument ceases to function, suspect a broken cable. Inspection will show whether the inner cable has broken; if so, the inner cable alone can be renewed and reinserted in the outer casing, after greasing. Never fit a new inner cable alone if the outer covering is damaged or compressed at any point.

20 Speedometer and tachometer drive: location and examination

1   The speedometer drive gearbox is fitted on the front wheel spindle where it is driven internally by the left-hand side of the wheel hub.
2   The gearbox rarely gives trouble if it is lubricated with grease at regular intervals. This can only be done after the wheel has been removed and the gearbox has been detached since no external grease nipple is fitted. The gearbox can be pulled from

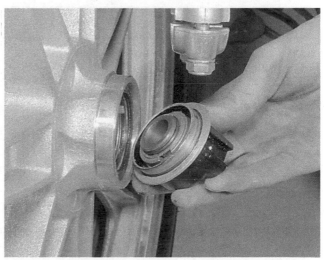

20.1a Speedometer drive gearbox is a push fit in hub and ...

20.1b ... engages with dog plate retained by oil seal

20.1c Boss on speedometer must locate with projection on fork leg

position after wheel removal.

3   If wear in the drive mechanism occurs, the worm shaft can be withdrawn after pulling out the threaded boss, which is a push fit. The drive pinion is retained to the inside of the brake plate by a circlip, in front of the shaped driving plate that takes up the drive from the wheel hub.

4   The tachometer drive is taken from the gear pinion, integral with the clutch outer drum, via the kickstart idler pinion. It is unlikely that the drive will give trouble during normal service life of the machine.

21 Cleaning the machine

1   After removing all surface dirt with a rag or sponge washed frequently in clean water, the machine should be allowed to dry thoroughly. Application of car polish or wax to the cycle parts will give a good finish, particularly if the machine has been neglected for a long period.

2   The plated parts of the machine should require only a wipe with a damp rag. If the plated parts are badly corroded, as may

occur during the winter when the roads are salted, it is preferable to use one of the proprietary chrome cleaners. These often have an oil base, which will help to prevent the corrosion from recurring.

3   If the engine parts are particularly oily, use a cleaning compound such as 'Gunk' or 'Jizer'. Apply the compound whilst the parts are dry and work it in with a brush so that it has the opportunity to penetrate the film of grease and oil. Finish off by washing down liberally with plenty of water, taking care that it does not enter the carburettors or the electrics. If desired, the now clean aluminium alloy parts can be enhanced further by using a special polish such as Solvol 'Autosol'', which will fully restore their brilliance.

4   Whenever possible, the machine should be wiped down after it has been used in the wet, so that it is not garaged under damp conditions which will promote rusting. Make sure to wipe the chain and re-oil it, to prevent water from entering the rollers and causing harshness with an accompanying high rate of wear. Remember there is little chance of water entering the control cables and causing stiffness of operation if they are lubricated regularly as recommended in the Routine Maintenance Section.

22 Fault diagnosis: frame and forks

Symptom	Cause	Remedy
Machine veers either to the left or the right with hands off handlebars	Bent frame Twisted forks Wheels out of alignment	Check and if necessary, renew. Check and if necessary, renew. Check and realign.
Machine rolls at low speed	Overtight steering head bearings	Slacken until adjustment is correct.
Machine judders when front brake is applied	Slack steering head bearings Worn fork bushes	Tighten, until adjustment is correct. Dismantle forks and bushes.
Machine pitches on uneven surfaces	Ineffective fork dampers Ineffective rear suspension units Suspension too soft	Check oil content. Check whether units still have damping action. Raise suspension unit adjustment one notch.
Fork action stiff	Fork legs out of alignment (twisted in yokes)	Slacken yoke clamps and fork top bolts. Pump fork several times then retighten from bottom upwards.
Machine wanders. Steering imprecise. Rear wheel tends to hop	Worn swinging arm pivot	Dismantle and renew bushes and pivot shaft.

Chapter 5 Wheels, brakes and tyres

Contents

Specifications

Tyres

Front	3.25 - 18 in 4PR (3.00 - 18 in spoked wheel)
Rear	3.50 - 18 in 4PR
Tyre pressures:	
Front	26 psi (1.8 kg/cm^2); * 20 psi (2.0 kg/cm^2)
Rear	28 psi (2.0 kg/cm^2); * 33 psi (2.0 kg/cm^2)

* For continuous high speed riding, or when a pillion passenger is carried.

Brakes

Front and rear:	
Type	Hydraulic, single disc
Disc diameter	267 mm (10.5 in)
Disc thickness	7.0 mm (0.275 in)
Service limit	6.5 mm (0.255 in)
Pad size	9.0 mm (0.354 in)
Service limit	4.5 mm (0.177 in)
Brake fluid specification	DOT 3 or SAE J1703a, b or c

1 General description

1 Both wheels are of 18 in diameter, the front carrying a ribbed tread tyre and the rear a tyre with a block tread pattern. On all models the rear tyre has a 3.50 in section. The front tyre is either 3.00 in or 3.25 in depending on the place of original delivery, and whether spoked on cast alloy wheels are fitted. Early RD400 models and many RD400C models employ chromed steel rims, which are laced to aluminium alloy hubs. Available as standard in some countries on the RD400C model are one piece seven spoke wheels cast in an aluminium alloy. This type of wheel may also be fitted as an optional extra.

2 All models employ hydraulically operated disc brakes on both wheels. The single disc is attached to the right-hand side of the hub.

2 Front wheel: examination and renovation (spoked wheel models)

1 Place the machine on the centre stand so that the front wheel is raised clear of the ground. Spin the wheel and check the rim alignment. Small irregularities can be corrected by tightening the spokes in the affected area, although a certain amount of practice is necessary to prevent over-correction. Any flats in the wheel rim should be evident at the same time. These are more difficult to remove and in most cases it will be necessary to have the wheel rebuilt on a new rim. Apart from the effect on stability, a flat will expose the tyre bead and walls to greater risk of damage.

2 Check for loose or broken spokes. Tapping the spokes is the best guide to tension. A loose spoke will produce a quite different sound and should be tightened by turning the nipple in

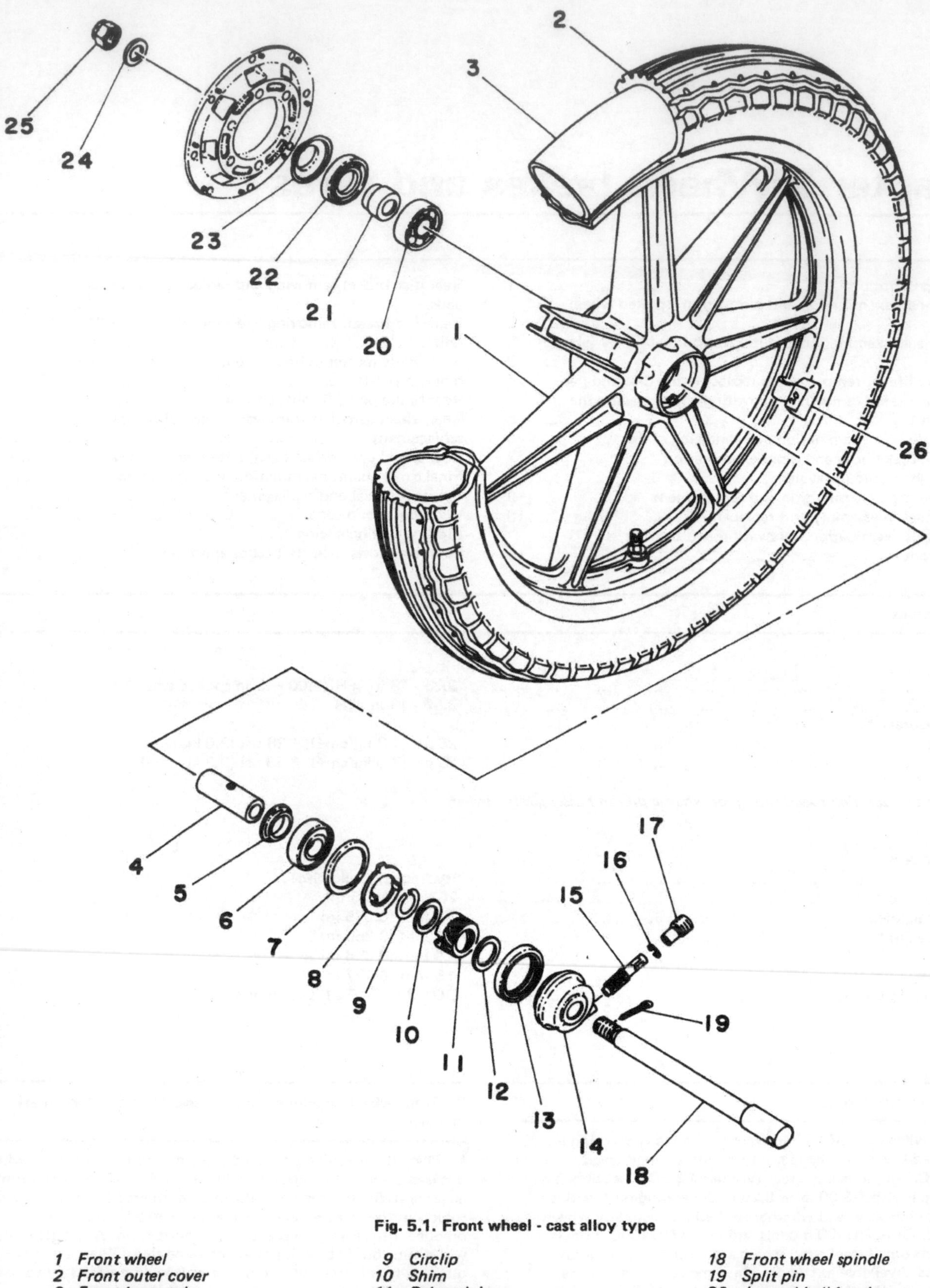

Fig. 5.1. Front wheel - cast alloy type

1 Front wheel	9 Circlip	18 Front wheel spindle
2 Front outer cover	10 Shim	19 Split pin
3 Front inner tube	11 Drive pinion	20 Journal ball bearing
4 Bearing spacer	12 Shim	21 Spacer
5 Bearing spacer flange	13 Oil seal	22 Oil seal
6 Sealed bearing	14 Speedometer drive gearbox	23 Dust cover
7 Backing ring	15 Driven shaft	24 Washer
8 Speedometer gearbox drive dog	16 Circlip	25 Castellated nut
	17 Gland union	26 Wheel balance weight

an anti-clockwise direction. Always re-check for run-out by spinning the wheel again. If the spokes have to be tightened an excessive amount, it is advisable to remove the tyre and tube by the procedure detailed in Section 19 of this Chapter; this is so that the protruding ends of the spokes can be ground off, to prevent them from chafing the inner tube and causing punctures.

3 Front wheel: examination and renovation (cast alloy wheel models)

1 Carefully check the complete wheel for cracks and chipping, particularly at the spoke roots and the edge of the rim. As a general rule a damaged wheel must be renewed as cracks will cause stress points which may lead to sudden failure under heavy load. Small nicks may be radiused carefully with a fine file and emery paper (No. 600 - No. 1000) to relieve the stress. If there is any doubt as to the condition of a wheel, advice should be sought from a Yamaha repair specialist.

2 Each wheel is covered with a coating of lacquer, to prevent corrosion. If damage occurs to the wheel and the lacquer finish is penetrated, the bared aluminium alloy will soon start to corrode. A whitish grey oxide will form over the damaged area, which in itself is a protective coating. This deposit however, should be removed carefully as soon as possible and a new protective coating of lacquer applied.

3 Check the lateral run out at the rim by spinning the wheel and placing a fixed pointer close to the rim edge. If the maximum run out is greater than 1.0 mm (0.03937 in), Yamaha recommend that the wheel be renewed. This is, however, a council of perfection; a run out somewhat greater than this can probably be accommodated without noticeable effect on steering. No means is available for straightening a warped wheel without resorting to the expense of having the wheel skimmed on all faces. If warpage was caused by impact during an accident, the safest measure is to renew the wheel complete. Worn wheel bearings may cause rim run out. These should be renewed as described in Section 9 of this Chapter.·.

4 Front disc brake: removing and replacing the disc and pads

1 The brake disc, attached to the right-hand side of the front wheel hub by eight bolts, rarely requires attention. Check for run out, which may have occurred as the result of crash damage, and for wear. Run out should not exceed 0.15 mm (0.006 in) at any point and the disc itself must not be permitted to wear below the limit thickness of 6.5 mm (0.255 in). If these figures are exceeded in either case, the disc must be renewed.

2 The disc bolts to a disc bracket, which itself bolts to the right-hand side of the wheel hub. It is necessary to detach the front wheel from the machine by first placing the machine on the centre stand so that the front wheel is raised clear of the ground. Slacken the clamp bolt at the base of the lower, left-hand fork leg and withdraw the split pin from the front wheel spindle nut, in this case on the right-hand side of the machine. Disconnect the speedometer drive cable. Slacken and remove the spindle nut, then withdraw the spindle; the wheel can now be withdrawn from the forks. It is preferable to insert a wooden wedge between the brake pads at this stage, to prevent them from being expelled if the brake lever is inadvertently operated. Remove the eight bolts from the inner edge of the brake disc and remove the disc. Note that each bolt is fitted with a lock washer, which must not be omitted during reassembly.

3 The two brake pads have an 'ear' moulded on the side, which locates with a slot in the caliper. To remove the pads, insert a screwdriver into the slot and prise each pad out of position.

4 The pads are moulded from a special resin impregnated asbestos compound and if renewal is necessary, only the correct replacement should be fitted. Each pad has an overall thickness of 9.0 mm (0.354 inches) when new and has a red line painted around the periphery, which represents the wear limit. The wear limit is 4.5 mm (0.177 inches) and on no account should the

4.2a Disc retained on bracket by eight bolts and tab washers

4.2b Disc bracket held by six bolts and three tab plates

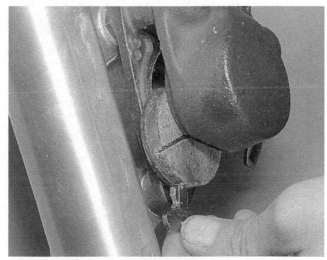

4.4a Disc pads can be removed with ease

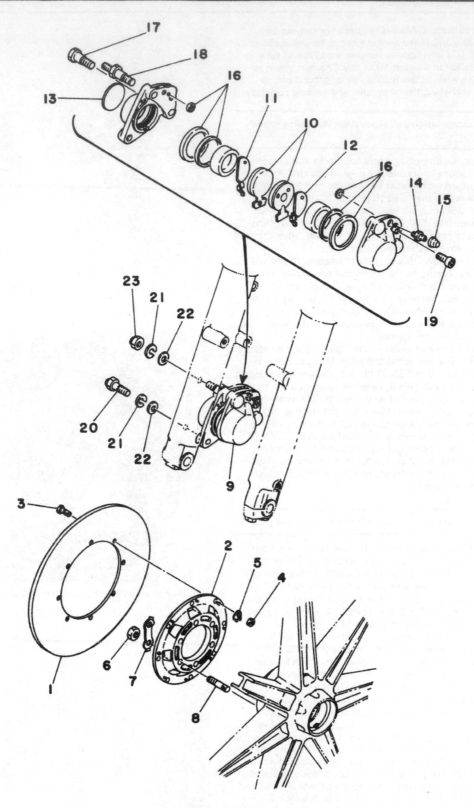

Fig. 5.2. Front disc brake caliper assembly

1	Brake disc	8	Stud - 6 off	16	Caliper seal set
2	Disc bracket	9	Caliper assembly - complete	17	Retaining bolt
3	Bolt - 8 off	10	Brake pad - 2 off	18	Special retaining bolt
4	Nut - 8 off	11	RH anti-chatter plate	19	Socket screw - 2 off
5	Tab washer - 8 off	12	LH anti-chatter plate	20	Bolt
6	Nut - 6 off	13	Emblem	21	Spring washer - 2 off
7	Tab washer - 3 off	14	Bleed nipple	22	Plain washer - 2 off
		15	Dust cap	23	Dome nut

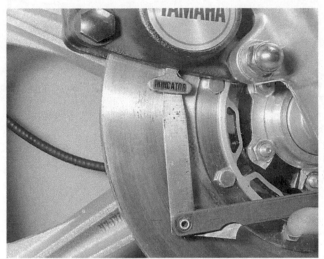

4.4b Pad wear should be checked with feeler gauge

4.5 DO NOT omit anti-chatter shims

pad be allowed to wear beyond an overall thickness of less than this amount. Each pad is fitted with a protruding ear marked 'indicator' which allows the condition of the pads to be ascertained without any need to remove them. Using a feeler gauge check the distance between the indicator 'ear' inside face and the face of the disc. If the distance is less than 0.5 mm (0.01969 inches) on either pad, both pads must be renewed. It is preferable to remove the front wheel before the pads are removed although the caliper mounting bolts can be slackened so that the unit can be swung clear of the disc as an alternative. Do NOT apply the brake in an attempt to displace the pads. If the actuating pistons move beyond their normal limit of travel, air will be admitted to the hydraulic system, necessitating a complete bleed of the system when reassembly is completed.

5   Reassembly is accomplished by reversing the procedure used for dismantling. The anti chatter shim that lies between each pad and the piston must be replaced with the straight edge facing forwards. Make sure that the brake pads are correctly located in the caliper and that the front wheel revolves quite freely when reassembly is complete. Always check the brake action before taking the machine on the road.

5   Front disc brake: removing, renovating and replacing the caliper unit

1   Before the caliper assembly can be removed from the right-hand fork leg, it is first necessary to drain off the hydraulic fluid. Disconnect the brake pipe at the union connection it makes with the caliper unit and allow the fluid to drain into a clean container. It is preferable to keep the front brake lever applied throughout this operation, to prevent the fluid from leaking out of the reservoir. A thick rubber band cut from a section of inner tube will suffice, if it is wrapped tightly around the lever and the handlebars.
2   Note that brake fluid is an extremely efficient paint stripper. Take care to keep it away from any paintwork on the machine or from any clear plastic, such as that sometimes used for instrument glasses.
3   When the fluid has drained off, remove the caliper mounting bolts and nuts, then rotate the caliper unit upwards and lift it away from the disc and the machine. The brake pads can now be removed, using a screwdriver to prise them from their respective housing. Detach the anti-chatter springs from each pad. Remove the two bolts that hold the two sections of the caliper unit together and when they have separated, the seal from the brake fluid inlet.
4   To displace the pistons, apply a blast of compressed air

through the brake fluid inlet of each caliper section. Take care to catch each piston as it emerges from its bore - if dropped or prised out of position with a screwdriver, it may be damaged irreparably and will have to be replaced. Remove the piston seal and dust seal from each caliper section.
5   The parts removed should be cleaned thoroughly, using only brake fluid as the liquid. Petrol, oil or paraffin will cause the various seals to swell and degrade, and should not be used under any circumstances. When the various parts have been cleaned, they should be stored in polythene bags until reassembly, so that they are kept dust free.
6   Examine the pistons for score marks or other imperfections. If they have any imperfections they must be renewed, otherwise air or hydraulic fluid leakage will occur, which will impair braking efficiency. With regard to the various seals, it is advisable to renew them all, irrespective of their appearance. It is a small price to pay against the risk of a sudden and complete front brake failure. It is standard Yamaha practice to renew the seals every two years, even if no braking problems have occurred.
7   Reassemble under clinically-clean conditions, by reversing the dismantling procedure. Renew the caliper unit bridge bolts as a safety precaution, even if they appear undamaged. Reconnect the hydraulic fluid pipe and make sure the union has been tightened fully. Before the brake can be used, the whole system must be bled of air, by following the procedure described in Section 8 of this Chapter.

6   Master cylinder: examination and renewing seals

1   The master cylinder and hydraulic fluid reservoir take the form of a combined unit mounted on the right-hand side of the handlebars, to which the front brake lever is attached.
2   Before the master cylinder unit can be removed and dismantled, the system must be drained. Place a clean container below the brake caliper unit and attach a plastic tube from the bleed screw of the caliper unit to the container. Lift off the master cylinder cover (cap), gasket and diaphragm, after removing the four countersunk retaining screws. Open the bleed screw one complete turn and drain the system by operating the brake lever until the master cylinder reservoir is empty. Close the bleed screw and remove the tube.
3   Before dismantling the master cylinder, it is essential that a clean working area is available on which the various component parts can be laid out. Use a sheet of white paper, so that none of the smaller parts can be overlooked.
4   Disconnect the stop lamp switch and front brake lever, taking care not to misplace the brake lever return spring. The stop lamp

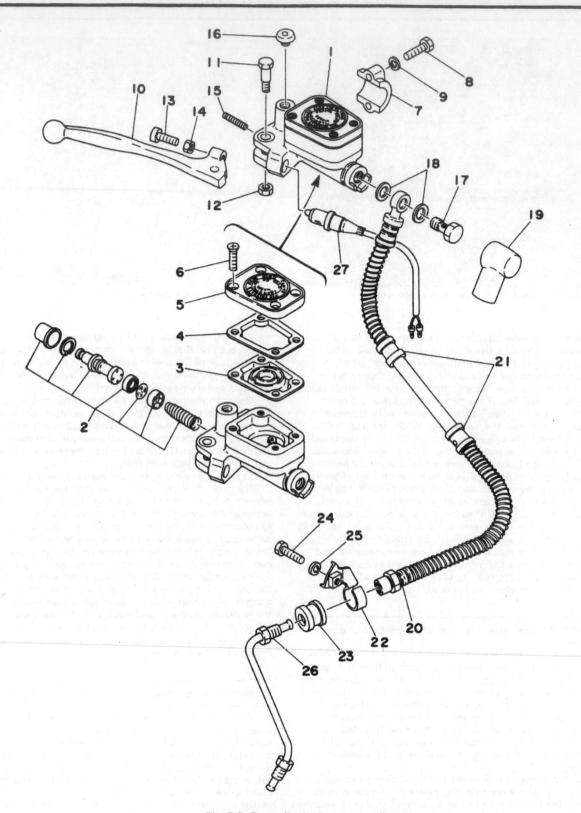

Fig. 5.3. Front disc brake master cylinder

1	Master cylinder	10	Front brake lever	19	Union boot
2	Cylinder piston assembly complete	11	Pivot bolt	20	Brake hose
3	Reservoir diaphragm	12	Nut	21	Clamp rubber - 2 off
4	Reservoir gasket	13	Adjuster bolt	22	Hose guide
5	Reservoir cap	14	Locknut	23	Grommet
6	Countersunk screw	15	Lever return spring	24	Bolt
7		16	Blind grommet	25	Spring washer
8	Bolt - 2 off	17	Banjo bolt	26	Brake pipe
9	Spring washer - 2 off	18	Sealing washer - 2 off	27	Front brake stoplamp switch

switch is a push fit in the lever stock. The lever pivots on a
bolt retained by a single nut. Remove the brake hose by unscrewing the banjo union bolt. Take the master cylinder away from the
handlebars by removing the two bolts that clamp it to the handlebars. Take care not to spill any hydraulic fluid on the paintwork
or on plastic or rubber components.

5   Withdraw the rubber boot that protects the end of the master
cylinder and remove the snap ring that holds the piston assembly
in position, using a pair of circlip pliers. The piston assembly can
now be drawn out, followed by the return valve, spring cup and
return spring.

6   The spring cup can now be separated from the end of the
return valve spring and the main cup prised off the piston.

7   Examine the piston and the cylinder cup very carefully. If
either is scratched or has the working surface impaired in any
other way, it must be renewed without question. Reject the
various seals, irrespective of their condition, and fit new ones in
their place. It often helps to soften them a little before they are
fitted by immersing them in a container of clean brake fluid.

8   When reassembling, follow the dismantling procedure in
reverse, but take great care that none of the component parts is
scratched or damaged in any way. Use brake fluid as the lubricant whilst reassembling. When assembly is complete, reconnect
the brake fluid pipe and tighten the banjo union bolt.
Use two new sealing washers at the union so that the banjo
bolt does not require overtightening to effect a good seal. Refill
the master cylinder with DOT 3 or SAE J1703 brake fluid and
bleed the system of air by following the procedure described in
Section 8 of this Chapter.

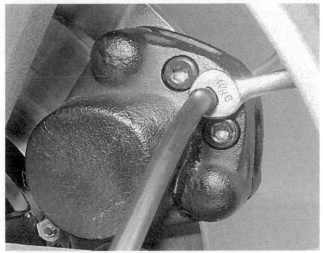
8.4a Use tight-fitting tubing on brake bleed nipple

7   Hydraulic brake hose and pipe: examination

1   An external brake hose and pipe is used to transmit the
hydraulic pressure to the caliper unit when the front brake or
rear brake is applied. The brake hose is of the flexible type, fitted
with an armoured surround. It is capable of withstanding pressures
up to 350 kg/cm^2. The brake pipe attached to it is made from
double steel tubing, zinc plated to give better corrosion
resistance.

2   When the brake assembly is being overhauled, check the
condition of both the hose and the pipe for signs of leakage or
scuffing, if either has made rubbing contact with the machine
whilst it is in motion. The union connections at either end must
also be in good condition, with no stripped threads or damaged
sealing washers. Check also the feed pipe from the rear brake
master cylinder to the reservoir. This pipe is not subjected to
pressure but may perish after a considerable length of time. It
is a push fit on the unions and is retained by screw clips.

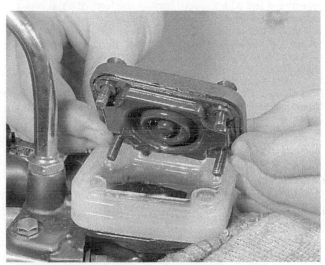
8.4b Front brake reservoir cap retained by four screws

8   Bleeding the hydraulic system

1   As mentioned earlier, brake action is impaired or even rendered
inoperative if air is introduced into the hydraulic system. This
can occur if the seals leak, the reservoir is allowed to run dry
or if the system is drained prior to the dismantling of any
component part of the system. Even when the system is refilled
with hydraulic fluid, air pockets will remain and because air will
compress, the hydraulic action is lost.

2   Check the fluid content of the reservoir and fill almost to the
top. Remember that hydraulic brake fluid is an excellent paint
stripper, so beware of spillage, especially near the petrol tank.

3   Place a clean glass jar below the brake caliper unit and attach
a clear plastic tube from the caliper bleed screw to the container.
Place some clean hydraulic fluid in the container so that the pipe
is always immersed below the surface of the fluid.

4   Unscrew the bleed screw one complete turn and pump the
handlebar lever slowly. As the fluid is ejected from the bleed
screw the level in the reservoir will fall. Take care that the level
does not drop too low whilst the operation continues, otherwise
air will re-enter the system, necessitating a fresh start.

5   Continue the pumping action with the lever until no further

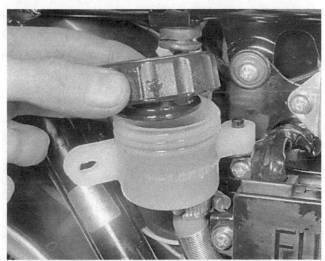
8.4c Rear brake reservoir pivots outwards for cap removal

air bubbles emerge from the end of the plastic pipe. Hold the
brake lever against the handlebars and tighten the caliper bleed
screw. Remove the plastic tube **AFTER** the bleed screw is
closed.
6   Check the brake action for sponginess, which usually denotes
there is still air in the system. If the action is spongy, continue
the bleeding operation in the same manner, until all traces of air
are removed.
7   Bring the reservoir up to the correct level of fluid and
replace the diaphragm, sealing gasket and cap. Check the entire
system for leaks. Recheck the brake action.
8   Note that fluid from the container placed below the brake
caliper unit whilst the system is bled, should not be re-used,
as it will have become aerated and may have absorbed moisture.
Allow the bled fluid to stand for at least 24 hours before use.

9   Wheel bearings: examination and replacement

1   Access to the front wheel bearings may be made after
removal of the wheel from the forks. Pull the speedometer
gearbox out of the hub left-hand boss and remove the dust seal
cover and wheel spacer from the hub right-hand side.
2   Lay the wheel on the ground with the disc side facing
downward and with a special tool, in the form of a rod with a
curved end, insert the curved end into the hole in the centre of
the spacer separating the two wheel bearings. If the other end
of the special tool is hit with a hammer, the right-hand bearing,
bearing flange washer, and bearing spacer will be expelled from
the hub.
3   Invert the wheel and drive out the left-hand bearing by
inserting a drift of the appropriate size, through the hub. During
the removal of either bearing it may be necessary to support the
wheel across an open-ended box so that there is sufficient clearance for the bearing to be displaced completely from the hub.
4   Remove all the old grease from the hub and bearings, giving
the latter a final wash in petrol. Check the bearings for signs of
play or roughness when they are turned. If there is any doubt
about the condition of a bearing, it should be renewed.
5   Before replacing the bearings, first pack the hub with new
grease. Then drive the bearings back into position, not forgetting
the distance piece that separates them. Take great care to ensure
that the bearings enter the housings perfectly squarely otherwise
the housing surface may be broached. Fit replacement oil seals
and any dust covers or spacers that were also displaced during
the original dismantling operation.

10   Front wheel: reassembly and replacement

1   Refit the speedometer gearbox to the left-hand side of the
hub, ensuring that the drive dogs engage correctly.
2   Lift the wheel upwards so that the brake disc enters the
brake caliper squarely and align the wheel so that the front wheel
spindle can be inserted from the left. Push the spindle home
through the right-hand fork leg until the head of the spindle is
flush with the left-hand fork leg. Ensure that the axial slot on the
speedometer gearbox is located with the projecting lug on the
inside of the left-hand fork lower leg. This is most important
otherwise the speedometer gearbox may revolve with the wheel
and snap the drive cable. Replace the washer and castellated nut
on the end of the fork spindle and tighten the nut. Replace the
split pin that passes through the end of the nut and the spindle.
Re-tighten the split clamp at the extreme end of the right-hand
fork leg, by tightening the forward nut first, followed by the rear
nut. A gap will remain between the rear face of the clamp and
the bottom rear face of the fork leg.
3   Spin the wheel to ensure that it moves freely, then attach the
speedometer drive cable, which is retained by a knurled ring.
Make sure that the cable passes through the cable retaining loop
of the mudguard, to ensure it cannot come into contact with
either the tyre or wheel.

11   Rear wheel: examination, renovation and bearing replacement

1   When inspecting the rear wheel follow the procedure given
for front wheel inspection in either Section 2 or 3 of this
Chapter, depending upon whether the wheel is of the spoked or
cast alloy type.
2   The procedure for removal, examination and lubrication of
the rear wheel bearings is materially the same as that used when
attending to the front wheel bearings. Follow the procedure
given in Section 9 of this Chapter.

12   Rear disc brake: removing and replacing the disc and pads

1   The rear brake disc and disc bracket are identical to the
components fitted to the front wheel. Removal and examination
of the disc and also the brake pads may be carried out by referring
to the information given in Section 4 of this Chapter.

8.7a Do not allow fluid to drop below level line

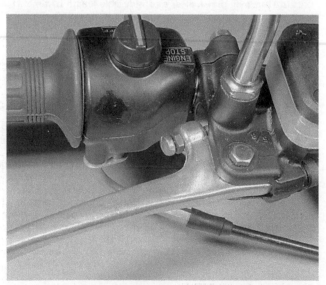

8.7b Free play at front brake lever may be adjusted by screw

9.1 Remove wheel spacers before drifting out bearings

9.5a Note fitted spacer on sprocket side rear wheel bearing

9.5b DO NOT omit bearing spacers on reassembly

9.5c Bearings may be tapped home using suitable socket ...

9.5d ... as can all oil seals

9.5e Left-hand front wheel oil seal retains speedometer drive dog

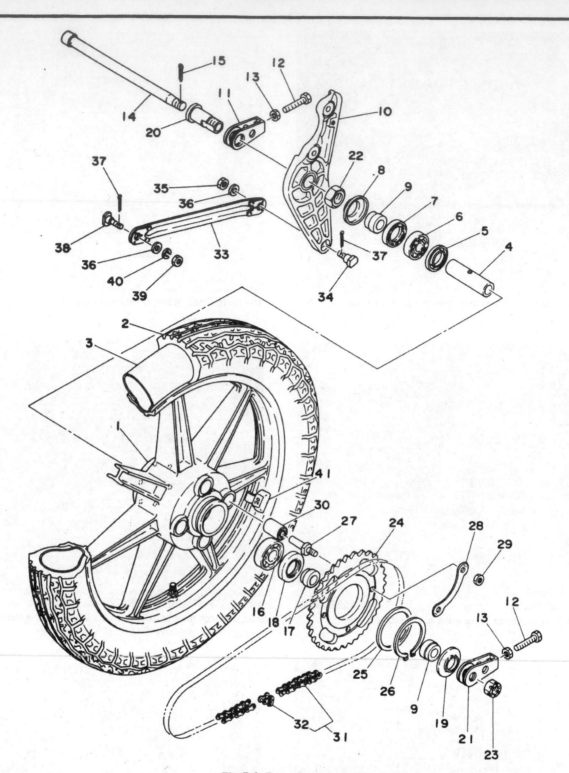

Fig. 5.4. Rear wheel - cast alloy type

1 Rear wheel	15 Split pin	29 Nut - 4 off
2 Rear outer cover	16 Journal ball bearing	30 'Flexibloc' rubber - 4 off
3 Rear inner tube	17 Spacer	31 Final drive chain
4 Bearing spacer	18 Oil seal	32 Master link
5 Bearing spacer flange	19 Dust cover	33 Brake torque arm
6 Journal ball bearing	20 Hollow bolt	34 Special bolt
7 Oil seal	21 LH chain adjuster	35 Nut
8 Dust cover	22 Nut	36 Plain washer - 2 off
9 Spacer - 2 off	23 Castellated nut	37 Split pin - 2 off
10 Caliper bracket	24 Final drive sprocket (38T)	38 Special bolt
11 RH chain adjuster	25 Plain washer	39 Nut
12 Bolt chain adjuster - 2 off	26 Circlip	40 Spring washer
13 Locknut - 2 off	27 Cush drive pin - 4 off	41 Wheel balance weight - A/R
14 Rear wheel spindle	28 Tab washer - 2 off	

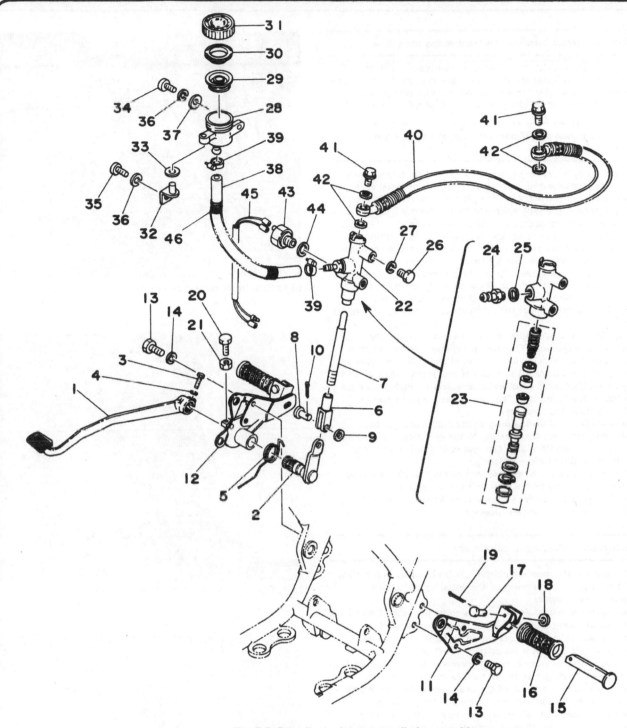

Fig. 5.5. Rear disc brake master cylinder assembly

1 Rear brake pedal	16 Footrest rubber - 2 off	31 Reservoir cap
2 Pivot shaft	17 Clevis pin - 2 off	32 Pivot/bracket
3 Pinch bolt	18 Plain washer - 2 off	33 Plain washer
4 Spring washer	19 Split pin - 2 off	34 Screw
5 Pedal return spring	20 Pedal height adjuster bolt	35 Screw
6 Clevis	21 Locknut	36 Spring washer - 2 off
7 Pushrod	22 Rear master cylinder	37 Plain washer
8 Clevis pin	23 Cylinder piston seal set	38 Reservoir hose
9 Plain washer	24 Feed union	39 Hose clip - 2 off
10 Split pin	25 Sealing washer	40 Brake hose
11 LH footrest bracket	26 Bolt - 2 off	41 Banjo bolt - 2 off
12 RH footrest/pedal bracket	27 Spring washer - 2 off	42 Plain washer - 4 off
13 Bolt - 4 off	28 Hydraulic fluid reservoir	43 Rear brake stoplamp switch
14 Spring washer - 4 off	29 Reservoir diaphragm	44 Plain washer
15 Footrest peg - 2 off	30 Reservoir gasket	45 Electrical lead
		46 Anti-kink spring

13 Rear disc brake: removing and renovating the caliper

The caliper unit fitted to the rear brake is similar in all respects to that utilised on the front brake system. Removal and inspection is materially the same and should be carried out by following the procedure given in Section 5 of this Chapter.

14 Rear brake master cylinder: removal, examination and renewing seals

1　The rear brake master cylinder is attached to the right-hand rear frame downtube and is operated from a foot pedal via a push rod connected to the pedal by a clevis fork and pin. The master cylinder reservoir is a separate component, remote from the master cylinder and interconnected by a short feed hose.
2　Drain the master cylinder and reservoir, using a similar technique to that described for the front brake master cylinder. The master cylinder reservoir is fitted with a screw cap.
3　Loosen the lower screw clip on the hydraulic fluid feed pipe and pull the pipe off the union. Take care not to drop any remaining fluid on the paint work. Unscrew the reservoir retaining bolt and then lift the reservoir upwards off the support bracket.
4　Disconnect the hydraulic hose at the brake caliper by removing the banjo bolt. The master cylinder is retained on the frame lug by two bolts. After removal of the bolt, the cylinder unit may be lifted upwards so that the operating push rod leaves the cylinder. Disconnect the brake lamp leads which are a push fit on the switch terminals. The master cylinder can now be lifted away from the machine.
5　Examination and dismantling of the rear brake master cylinder may be made by referring to the directions in Section 5 of this Chapter. Additionally, the reservoir should be flushed out with clean fluid before refitting.
6　After reassembly and replacement of the rear brake master cylinder components which may be made by reversing the dismantling procedure - bleed the rear brake system of air by referring to Section 8 of this Chapter.

15 Rear brake pedal height: adjustment

1　The pivot shaft upon which the rear brake pedal is mounted is splined to allow adjustment of the pedal height to suit individual requirements.
2　To adjust the height, loosen and remove the pinch bolt which passes into the rear of the pedal box. Draw the pedal off the splines and refit it at the required angle. Ideally the pedal should be fitted, so that it is positioned just below the rider's right foot, when the rider is seated normally. In this way the foot does not have to be lifted before the brake can be applied.
3　The upper limit of travel of the brake pedal may be adjusted by means of the bolt and locknut fitted to the pedal pivot mounting bracket. Care should be exercised when lowering the pedal by this method as movement imparted to the master cylinder piston may actuate the brake to a small degree. Adjustment of the pedal may necessitate readjustment of the rear stop lamp switch.

16 Rear wheel sprocket: examination and replacement

1　The rear wheel sprocket is retained on the left-hand side of the hub by a large circlip, and is located by four pegs which pass into the cush drive in the hub and are retained by hexagonal nuts.
2　To remove the sprocket, detach the large circlip and the spacing plate which lies below. The sprocket can be removed complete with drive pegs, but it is probable that the pegs are seized in the steel sleeves bonded to the flexible rubbers. If this is the case, knock down the 'ears' of the tab washers and remove

14.4a Rear brake master cylinder is retained by two bolts

14.4b Operating pushrod passes through boot into cylinder

16.1a Sprocket retained by large circlip which ...

16.1b ... lies below single plate acting as tab washer for ...

16.2 ... cush drive pins passing into flexible bushes

16.4 Replace sprocket so that flats on pins engage correctly

18.1a Torque arm bolt must be slackened as must ...

the peg securing nuts. The sprocket can then be lifted off with ease.

3　Check the condition of the sprocket teeth. If they are hooked, chipped or badly worn, the sprocket must be renewed. It is considered bad practice to renew one sprocket on its own. The final drive sprockets should always be renewed as a pair and a new chain fitted, otherwise rapid wear will necessitate even earlier renewal on the next occasion.

4　The sprocket may be refitted by reversing the dismantling procedure. It is important that the recesses in the rear of the sprocket are engaged correctly by the milled flats on each cush drive pin.

17 Rear wheel cush drive: examination and renovation

1　The cush drive assembly consists of four tubular rubber bushes located in the hub. The four special pegs retained by nuts on the sprocket locate with these bushes, to give a cushioning effect to the sprocket and drive.

2　To obtain access to the bushes, the sprocket has to be removed by detaching its circlip and pulling it from the wheel hub. Renewal of the bushes is required when there is excessive

sprocket movement. As stated in the previous Section, the pegs may seize after a considerable length of time. If this occurs, the sprocket complete with pegs should be drawn from the hub using a sprocket puller. A blanking plate or bar, fabricated from mild steel, will have to be made and placed over the bearing. The sprocket puller screw can then bear against the bar.

3　Removal of the flexible bushes is almost impossible without the use of a special expanding extractor. It is recommended that the wheel be returned to a Yamaha dealer who can carry out the work without risk of damage to the wheel.

18 Final drive chain: examination and lubrication

1　The final drive chain is fully exposed, with only a light chainguard over the top run. Periodically the tension will need to be adjusted, to compensate for wear. This is accomplished by placing the machine on the centre stand and slackening the wheel nuts on the left-hand side of the rear wheel so that the wheel can be drawn backward by means of the drawbolt adjusters in the fork ends. The rear brake torque arm bolt and the caliper bracket nut must also be slackened during the operation.

2　The chain is in correct tension if there is approximately 20 mm (¾ inch) slack in the middle of the lower run. Always check

18.1b ... rear caliper bracket hollow bolt to adjust chain

when the chain is at its tightest point as a chain rarely wears evenly during service.

3   Always adjust the drawbolts an equal amount in order to preserve wheel alignment. The fork ends are clearly marked with a series of vertical lines above the adjusters, to provide a simple, visual check. If desired, wheel alignment can be checked by running a plank of wood parallel to the machine, so that it touches the side of the rear tyre. If the wheel alignment is correct, the plank will be equidistant from each side of the front wheel tyre, when tested on both sides of the rear wheel. It will not touch the front wheel tyre because this tyre is of smaller cross section. See accompanying diagram.

4   Do not run the chain overtight to compensate for uneven wear. A tight chain will place undue stress on the gearbox and rear wheel bearings, leading to their early failure. It will also absorb a surprising amount of power.

5   After a period of running, the chain will require lubrication. Lack of oil will greatly accelerate the rate of wear of both the chain and the sprockets and will lead to harsh transmission. The application of engine oil will act as a temporary expedient, but it is preferable to remove the chain and clean it in a paraffin bath before it is immersed in a molten lubricant such as "Linklyfe" or "Chainguard". These lubricants achieve better penetration of the chain links and rollers and are less likely to be thrown off when

18.3 Index lines on fork ends aid wheel alignment

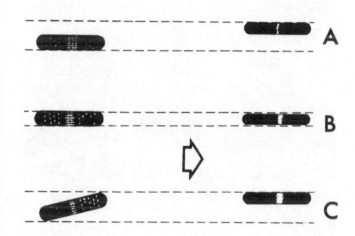

Fig. 5.6. Method of checking wheel alignment

A & C Incorrect;   B   Correct

18.5 Aerosol lubricant may be used for intermediate lubrication

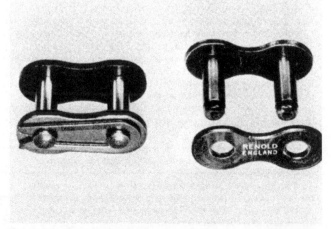

18.8 Chains of British manufacture are available as replacements

Tyre changing sequence - tubed tyres

 A Deflate tyre. After pushing tyre beads away from rim flanges push tyre bead into well of rim at point opposite valve. Insert tyre lever adjacent to valve and work bead over edge of rim.

Use two levers to work bead over edge of rim. Note use of rim protectors **B**

 C Remove inner tube from tyre

When first bead is clear, remove tyre as shown **D**

 E When fitting, partially inflate inner tube and insert in tyre

Work first bead over rim and feed valve through hole in rim. Partially screw on retaining nut to hold valve in place. **F**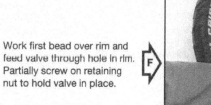

 G Check that inner tube is positioned correctly and work second bead over rim using tyre levers. Start at a point opposite valve.

Work final area of bead over rim whilst pushing valve inwards to ensure that inner tube is not trapped **H**

the chain is in motion.

6   To check whether the chain is due for replacement, lay it lengthwise in a straight line and compress it endwise so that all the play is taken up. Anchor one end and measure the length. Now pull the chain with one end anchored firmly, so that the chain is fully extended by the amount of play in the opposite direction. If there is a difference of more than ¼ inch per foot in the two measurements, the chain should be replaced in conjunction with the sprockets. Note that this check should be made AFTER the chain has been washed out, but BEFORE any lubricant is applied, otherwise the lubricant may take up some of the play.

7   When replacing the chain, make sure that the spring link is seated correctly, with the closed end facing the direction of travel.

8   Replacement chains are now available in standard metric sizes from Renold Limited, the British chain manufacturer. When ordering a new chain, always quote the size, the number of chain links and the type of machine to which the chain is to be fitted.

19 Tyres: removal and replacement

1   At some time or other the need will arise to remove and replace the tyres, either as the result of a puncture or because a renewal is required to offset wear. To the inexperienced, tyre changing represents a formidable task yet if a few simple rules are observed and the technique learned, the whole operation is surprisingly simple.

2   To remove the tyre from the wheel, first detach the wheel from the machine by following the procedure in Chapter 4, Sections 2.5 and 2.6 for the front wheel and Section 10.3 for the rear wheel. Deflate the tyre by removing the valve insert and when it is fully deflated, push the bead of the tyre away from the wheel rim on both sides so that the bead enters the centre well of the rim. Remove the locking cap and push the tyre valve into the tyre itself.

3   Insert a tyre lever close to the valve and lever the edge of the tyre over the outside of the wheel rim. Very little force should be necessary; if resistance is encountered it is probably due to the fact that the tyre beads have not entered the well of the wheel rim all the way round the tyre.

4   Once the tyre has been edged over the wheel rim, it is easy to work around the wheel rim so that the tyre is completely free on one side. At this stage, the inner tube can be removed.

5   Working from the other side of the wheel ease the other edge of the tyre over the outside of the wheel rim furthest away. Continue to work around the rim until the tyre is free from the rim.

6   If a puncture has necessitated the removal of the tyre, reinflate the inner tube and immerse it in a bowl of water to trace the source of the leak. Mark its position and deflate the tube. Dry the tube and clean the area around the puncture with a petrol-soaked rag. When the surface has dried, apply rubber solution and allow this to dry before removing the backing from the patch and applying the patch to the surface.

7   It is best to use a patch of the self-vulcanising type, which will form a very permanent repair. Note that it may be necessary to remove a protective covering from the top surface of the patch, after it has sealed in position. Inner tubes made from synthetic rubber may require a special type of patch and adhesive, if a satisfactory bond is to be achieved.

8   Before replacing the tyre, check the inside to make sure the agent that caused the puncture is not trapped. Check also the outside of the tyre, particularly the tread area, to make sure nothing is trapped that may cause a further puncture.

9   If the inner tube has been patched on a number of past occasions, or if there is a tear or large hole, it is preferable to discard it and fit a new one. Sudden deflation may cause an accident, particularly if it occurs with the front wheel.

10 To replace the tyre, inflate the inner tube sufficiently for it to assume a circular shape but only just. Then push it into the tyre so that it is enclosed completely. Lay the tyre on the wheel at an angle and insert the valve through the rim tape and the hole in the wheel rim. Attach the locking cap on the first few threads, sufficient to hold the valve captive in its correct location.

11 Starting at the point furthest from the valve, push the tyre bead over the edge of the wheel rim until it is located in the central well. Continue to work around the tyre in this fashion until the whole of one side of the tyre is on the rim. It may be necessary to use a tyre lever during the final stages.

12 Make sure there is no pull on the tyre valve and again commencing with the area furthest from the valve, ease the other bead of the tyre over the edge of the rim. Finish with the area close to the valve, pushing the valve up into the tyre until the locking cap touches the rim. This will ensure the inner tube is not trapped when the last section of the bead is edge over the rim with a tyre lever.

13 Check that the inner tube is not trapped at any point. Reinflate the inner tube, and check that the tyre is seating correctly around the wheel rim. There should be a thin rib moulded around the wall of the tyre on both sides, which should be equidistant from the wheel rim at all points. If the tyre is unevenly located on the rim, try bouncing the wheel when the tyre is at the recommended pressure. It is probable that one of the beads has not pulled clear of the centre well.

14 Always run the tyres at the recommended pressures and never under or over-inflate. The correct pressures for solo use are given in the Specifications section of this Chapter.

15 Tyre replacement is aided by dusting the side walls, particularly in the vicinity of the beads, with a liberal coating of french chalk. Washing-up liquid can also be used to good effect, but this has the disadvantage of causing the inner surfaces of the wheel rim to rust.

16 Never replace the inner tube and tyre without the rim tape in position. If this precaution is overlooked there is a good chance of the ends of the spoke nipples chafing the inner tube and causing a crop of punctures.

17 Never fit a tyre that has a damaged tread or side walls. Apart from the legal aspects, there is a very great risk of blow-out, which can have serious consequences on any two-wheeled vehicle.

18 Tyre valves rarely give trouble, but it is always advisable to check whether the valve itself is leaking before removing the tyre. Do not forget to fit the dust cap, which forms an effective second seal.

20 Valve cores and caps

1   Valve cores seldom give trouble, but do not last indefinitely. Dirt under the seating will cause a puzzling 'slow-puncture'. Check that they are not leaking by applying spittle to the end of the valve and watching for air bubbles.

2   A valve cap is a safety device, and should always be fitted. Apart from keeping dirt out of the valve, it provides a second seal in case of valve failure, and may prevent an accident resulting from sudden deflation.

21 Front wheel balancing

1   The front wheel should be statically balanced, complete with tyre. An out of balance wheel can produce dangerous wobbling at high speed.

2   Some tyres have a balance mark on the sidewall. This must be positioned adjacent to the valve. Even so, the wheel still requires balancing.

3 With the front wheel clear of the ground, spin the wheel several times. Each time, it will probably come to rest in the same position. Balance weights should be attached diametrically opposite the heavy spot, until the wheel will not come to rest in any set position, when spun.

4 Balance weights, which clip round the spokes, are available in 5, 10 or 20 gramme weight. If they are not available, wire solder wrapped round the spokes and secured with insulating tape will make a substitute.

5 It is possible to have a wheel dynamically balanced at some dealers. This requires its removal.

6 There is no need to balance the rear wheel under normal road conditions, although any tyre balance mark should be aligned with the valve.

7 Machines fitted with cast aluminium wheels require special balancing weights which are designed to clip onto the centre rim flange, much in the way that weights are affixed to car wheels. When fitting these weights, take care not to affix any weight nearer than 40 mm (1.54 in) to the radial centre line of any spoke. Refer to the accompanying diagram.

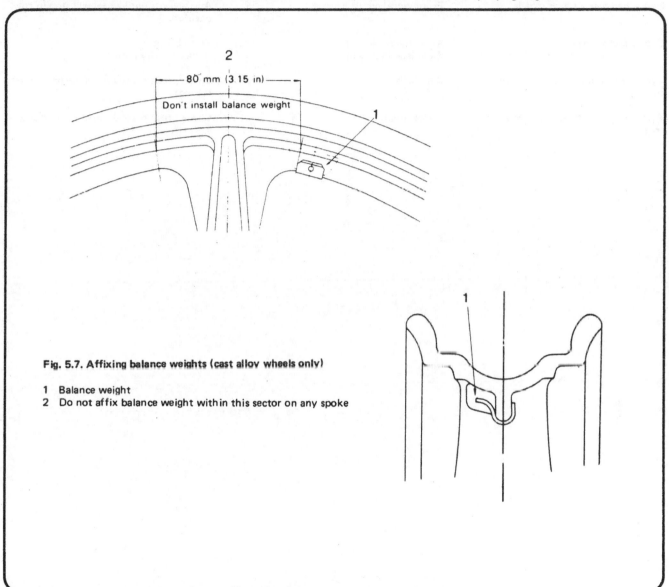

Fig. 5.7. Affixing balance weights (cast alloy wheels only)

1 Balance weight
2 Do not affix balance weight within this sector on any spoke

See next page for Fault diagnosis - wheels, brakes and tyres

22 Fault diagnosis: wheels, brakes and tyres

Symptom	Cause	Remedy
Handlebars oscillate at low speeds	Buckled front wheel	Remove wheel for specialist attention. Renew wheel (cast alloy type).
	Incorrectly fitted front tyre	Check whether line around bead is equidistant from rim.
Forks 'hammer' at high speeds	Front wheel out of balance	Add weights until wheel will stop in any position.
Brakes feel spongy	Air in hydraulic line	Bleed brakes.
	Fluid leak in system	Replace faulty part.
Tyres wear more rapidly in middle of tread	Over-inflation	Check pressures and run at recommended settings.
Tyres wear rapidly at outer edges of tread	Under-inflation	Check pressures and run at recommended settings.

Chapter 6 Electrical system

Contents

Specifications

Battery			UK	USA
Type	... ...		Lead acid	Lead acid
Make	... ...		Yuasa or Furukawa	Yuasa or Furukawa
Voltage	... ...		AYT2 - 12	AYT2 - 12
Capacity	... ...		5.5 amp hr	5.5 amp hr
Alternator				
Make	... ...		Hitachi	Mitsubishi
Type	... ...		LD118 - 02	AZ 2015Y
Voltage	... ...		12v	12v
Output	... ...		252 watts	280 watts
Regulator unit				
Make	... ...		Hitachi	Mitsubishi
Type	... ...		TR12 - 29	RFT 12M$_2$
Regulating voltage	... ...		14.5 volts	14.5 volts
Bulbs				
Headlamp	... ...		35/25 w pre focus	40/30 w pre focus
Pilot lamp	... ...		3.4 w bayonet fitting	3.4 w bayonet fitting
Tail/Stop lamp	... ...		5/21 w offset pin (2)	8/27 w offset pin (2)
Speedometer lamp	... ...		3.4 w bayonet fitting	3.4 w bayonet fitting
Tachometer lamp	... ...		3.4 w bayonet fitting	3.4 w bayonet fitting
Flasher indicator lamp	... ...		3.4 w bayonet fitting (2)	3.4 w bayonet fitting (2)
Neutral indicator	... ...		3.4 w bayonet fitting	3.4 w bayonet fitting
Flasher lamps	... ...		27 w each (4)	27 w each (4)
Headlamp flasher lamp	... ...		3.4 w bayonet fitting	3.4 w bayonet fitting
Oil level warning lamp	... ...		3.4 w bayonet fitting	3.4 w bayonet fitting

All bulbs rated 12 volt.

1 General description

1 The Yamaha RD 400 twin is fitted with a 12 volt electrical system. The circuit comprises a crankshaft-driven alternator, the output of which is controlled by a voltage regulator linked with the stator coil windings. Because the output from the alternator is ac a rectifier is included in the circuit to convert to dc in order to maintain the charge of the 12 volt, 5.5 amp hour battery.

2 Because the alternator has coils in the rotor assembly, brush gear is employed to pick up the current for the rotor from slip rings on the outer face of the rotor. No permanent magnets are employed in the construction of the alternator; it functions solely on the electro-magnetic principle.

2 Alternator: checking the output

1 As previously mentioned in Chapter 3.2, there is no satisfactory method of checking the output from the alternator without test equipment of the multi-meter type. If the performance of the alternator is in any way suspect, it should be checked by either a Yamaha repair specialist or auto-electrical mechanic.

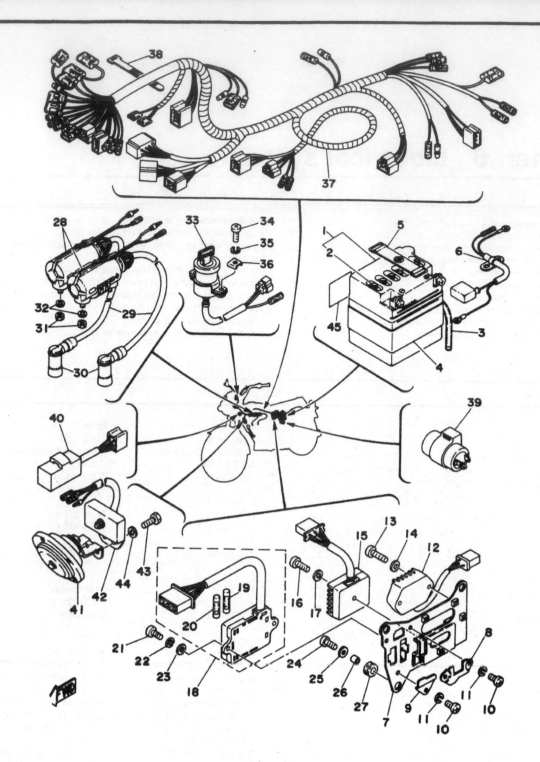

Fig. 6.1. Electrical equipment

1	Battery	16	Screw	31	Nut - 4 off
2	Cell plug - 6 off	17	Spring washer	32	Spring washer - 4 off
3	Breather hose	18	Fuse box	33	Ignition switch
4	Strap	19	10A fuse - 4 off	34	Screw - 2 off
5	Retaining strap	20	20A fuse - 2 off	35	Spring washer - 2 off
6	Main lead	21	Screw - 2 off	36	Special washer - 2 off
7	Component mounting plate	22	Spring washer - 2 off	37	Wiring harness
8	Secondary mounting plate A	23	Plain washer - 2 off	38	Lead strap - 2 off
9	Secondary mounting plate B	24	Screw - 3 off	39	Flasher unit
10	Screw - 3 off	25	Plain washer - 3 off	40	Flasher cancelling unit
11	Spring washer - 3 off	26	Spacer - 3 off	41	Horn
12	Voltage regulator	27	Grommet - 3 off	42	Resistor
13	Screw - 2 off	28	Ignition coil - 2 off	43	Bolt - 2 off
14	Spring washer - 2 off	29	HT lead - 2 off	44	Spring washer
15	Rectifier	30	Plug cap - 2 off	45	Battery label

3 Voltage regulator: location and checking

1 The voltage regulator is of the sealed and transistorised type and is located on the machine to the right of the battery, below the dualseat. It is the function of the voltage regulator to pass a controlled amount of voltage to the rotor windings so that the output from the alternator is matched to the requirements of the electrical system. By this means it is possible to keep the voltage supplied by the alternator within the 12-15 volt range so that there are no surges or sudden overloads.

2 If problems with the charging system have developed, but the alternator and rectifier have been tested and found to be in working order, the regulator may be checked using a dc volt-meter with a range of 0-20 volts. The battery should be fully charged for this test.

3 Connect the voltmeter across the battery terminals ie. with the positive lead connected to the positive battery terminal and the negative lead to the negative battery terminal. **Under no circumstances** should the battery be disconnected from the machine whilst the engine is running and the alternator charging. Start the engine and allow it to run at 2,000 rpm or more. The voltmeter should indicate a reading of 14 ± 0.3 volts. If the reading is substantially outside this range, the regulator should be replaced.

4 Battery: examination and maintenance

1 A Furukawa or Yuasa battery is fitted as standard. This battery is a lead-acid type and has a capacity of 5.5 amp hours.

2 The transparent plastic case of the battery permits the upper and lower levels of the electrolyte to be observed when the battery is lifted from its housing below the dualseat. Main-tenance is normally limited to keeping the electrolyte level between the prescribed upper and lower limits and by making sure the vent pipe is not blocked. The lead plates and their separators can be seen through the transparent case, a further guide to the general condition of the battery.

3 Unless acid is spilt, as may occur if the machine falls over, the electrolyte should always be topped up with distilled water, to restore the correct level. If acid is spilt on any of the machine, it should be neutralised with an alkali such as washing soda and washed away with plenty of water, otherwise serious corrosion will occur. Top up with sulphuric acid of the correct specific gravity (1.260 - 1.280) only when spillage has occurred. Check that the vent pipe is well clear of the frame tubes or any of the other cycle parts, for obvious reasons.

5 Battery: charging procedure

1 The normal charging rate for the 5.5 amp hour battery is 0.5 amps. A more rapid charge, not exceeding 1 amp can be given in an emergency. The higher charge rate should, if possible, be avoided since it will shorten the working life of the battery.

2 Make sure that the battery charger connections are correct, red to positive and black to negative. It is preferable to remove the battery from the machine whilst it is being charged and to remove the vent plug from each cell. When the battery is re-connected to the machine, the black lead must be connected to the negative terminal and the red lead to positive. This is most important, as the machine has a negative earth system. If the terminals are inadvertently reversed, the electrical system will be damaged permanently. The rectifier will be destroyed by a reversal of the current flow.

6 Rectifier: general description

1 This is a full wave rectifier composed of six silicon diodes. The diodes permit a one-way flow of electrical current and therefore convert the alternating current output from the alternator into direct current, which can be used for battery charging.

2 In the event of failure of the battery to maintain a fully-charged condition, it is possible that the rectifier is malfunctioning Unfortunately there is no easy way of checking without the appropriate electronic test equipment. Provided the electrical connections have not been inadvertently transposed at the battery, a check by substitution of the correct replacement is the only practicable method of verification.

7 Fuse: location and replacement

1 A bank of fuses is contained within a small plastic box located near the regulator and sharing the same mounting bracket. The box contains four 10A fuses and two 20A fuses, of which one of each type is spare.

2 Before replacing a fuse that has blown, check that no obvious short circuit has occurred, otherwise the replacement fuse will blow immediately it is inserted. It is always wise to check the electrical circuit thoroughly, to trace the fault and eliminate it.

3 When a fuse blows while the machine is running and no spare is available, a 'get you home' remedy is to remove the blown fuse and wrap it in silver paper before replacing it in the fuseholder. The silver paper will restore the electrical continuity by bridging the broken fuse wire. This expedient should NEVER be used if there is evidence of a short circuit or other major electrical fault, otherwise more serious damage will be caused. Replace the 'doctored' fuse at the earliest possible opportunity, to restore full circuit protection.

8 Headlamp: replacing bulbs and adjusting beam height

1 To remove the headlamp rim, detach the small screw on the right-hand underside of the headlamp shell. The rim can then be prised off, complete with the reflector unit.

2 The main bulb is a twin filament type, to give a dipped beam facility. The bulb holder is attached to the back of the reflector by a rubber sleeve, which fits around a flange in the reflector and the flange of the bulbholder. An indentation in the bulbholder orifice and a projection on the bulbholder ensures the bulb is always replaced in the same position so that the focus is unaltered.

3 It is not necessary to re-focus the headlamp when a new bulb is fitted. Apart from the just-mentioned method of location, the bulbs used are of the pre-focus type, built to a precise specification. To release the bulbholder, twist and lift away.

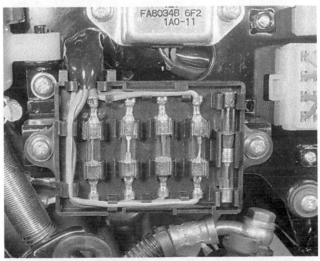

7.1 Fuses are contained in a bank, together with two spares

8.1 Headlamp rim/reflector retained by a single screw

8.2a Main bulb holder is located by rubber boot

8.2b Headlamp bulb is bayonet fixed by three offset pins

8.4 Bayonet fixed pilot bulb is also retained by rubber boot

10.1 Stop/tail lamp lens is retained by two screws

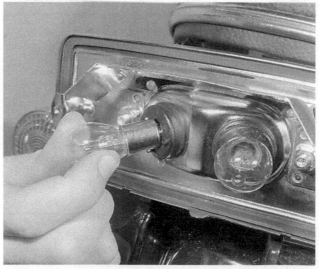

10.2 Two twin-filament bulbs are fitted

4 The pilot lamp bulbholder, like the bulb itself, has a bayonet fitting. It is protected by a rubber sleeve. Remove the bulbholder first, then the bulb.

5 The main headlamp bulb is rated at 35/25w, 12 volts and the pilot lamp bulb at 3w, 12 volts. Variations in the wattage may occur according to the country or state for which the machine is supplied. In the UK, the pilot bulb has a statutory minimum rating of 6w.

6 Beam alignment is adjusted by tilting the headlamp after the two retaining bolts have been slackened and then retightening them after the correct beam height is obtained, without moving the setting.

7 To obtain the correct beam height, place the machine on level ground facing a wall 25 feet distant, with the rider seated normally. The height of the beam centre should be equal to that of the height of the centre of the headlamp from the ground, when the dip switch is in the main beam position. Furthermore, the concentrated area of light should be centrally disposed. Adjustments in either direction are made by rearranging the angle of the headlamp, as described in the preceding paragraph. Note that a different beam setting will be needed when a pillion passenger is carried. If a pillion passenger is carried regularly, the passenger should be seated in addition to the rider when the beam setting adjustment is made.

9 The above instructions for beam setting relate to the requirements of the United Kingdom's transport lighting regulations. Other settings may be required in countries other than the UK.

9 Handlebar switches: function and replacement

1 The dipswitch forms part of the left-hand dummy twist grip which contains the horn button, flashing indicator lamp switch, and headlamp flasher. The right-hand twist grip assembly incorporates the lighting master switch and a three position ignition positive cut-out switch.

2 In the event of failure of any of these switches, the switch assembly must be replaced as a complete unit since it is not practicable to effect a permanent repair.

10 Stop and tail lamp: replacing the bulb

1 The tail lamp is fitted with two twin filament bulbs of 12 volt, 5/21w rating, to illuminate the rear number plate and rear of the machine, and to give visual warning when the rear brake is applied. To gain access to the bulbs remove the plastic lens cover, which is retained by two long screws. Check that the gasket between the lens cover and the main body of the lamp is in good condition.

2 Each bulb has a bayonet fitting and has staggered pins to prevent the bulb contacts from being reversed.

3 If the tail lamp bulbs keep blowing, suspect either vibration of the rear mudguard or more probably, an intermittent earth connection.

11 Flashing indicator lamps

1 The forward facing indicator lamps are connected to 'stalks' that replace the bolts on which the headlamp shell would normally be mounted. The stalks are hollow and have threaded ends so that they can be locked in position from the inside of the headlamp shell or to the fork lugs that carry the headlamp. The rear facing lamps are mounted on similar, shorter stalks, at a point immediately to the rear of the dualseat.

2 In each case, access to the bulb is gained by removing the plastic lens cover, which is retained by two screws. Bayonet fitting bulbs of the single filament type are used, each with a 12 volt 27w rating.

12 Flasher unit: location and replacement

1 The flasher relay unit is located either under the dualseat or behind the right-hand side cover that carries the capacity symbol of the model. It is retained by a single bolt passing through a built-in clip.

2 If the flasher unit is functioning correctly, a series of audible clicks will be heard when the indicator lamps are in action. If the unit malfunctions and all the bulbs are in working order, the usual symptom is one initial flash before the unit goes dead; it will be necessary to replace the unit complete if the fault cannot be attributed to any other cause.

3 In addition to the flasher unit, an electronic flasher cancelling unit is incorporated in the indicator system. The unit automatically turns the flasher light off a certain time after the flasher switch has been operated. The time lapse is dependent on the speed of the machine. If the machine is travelling fast, the unit cancels automatically after a short time. The slower the machine is travelling, the longer the time taken for cancellation. The system may be overidden manually in the normal manner.

3 Take great care when handling either unit because they are easily damaged if dropped.

11.2a Flasher lenses are retained by two screws

11.2b Bulbs are bayonet fixed with in line pins

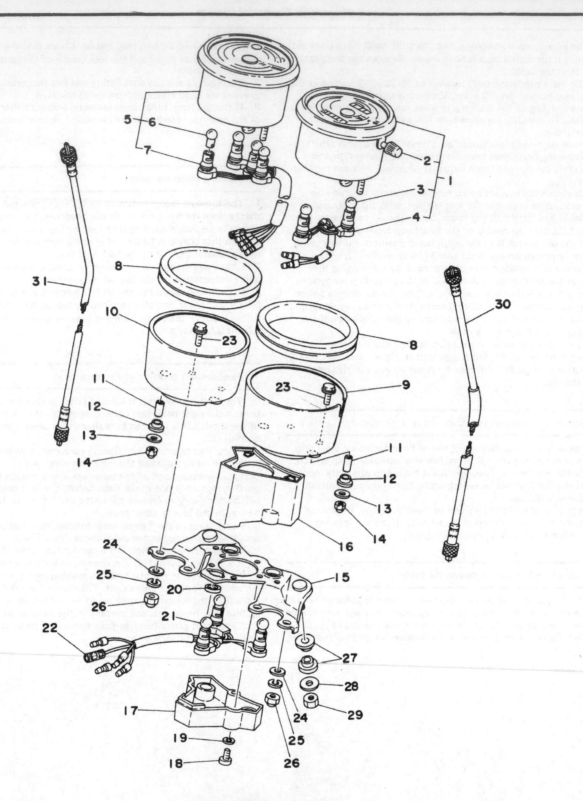

Fig. 6.2. Instrument assembly

1 Speedometer	11 Spacer - 4 off	21 Bulb - 3 off
2 Trip meter knob	12 Rubber insert - 4 off	22 3-bulb holder
3 Bulb - 2 off	13 Plain washer - 4 off	23 Bolt - 4 off
4 Double holder assembly	14 Dome nut - 4 off	24 Plain washer - 4 off
5 Tachometer	15 Instrument bracket	25 Spring washer - 4 off
6 Bulb - 4 off	16 Warning lamp console	26 Dome nut - 4 off
7 Four-bulb holder assembly	17 Warning lamp console base	27 Rubber insert - 4 off
8 Rubber seat - 2 off	18 Screw - 3 off	28 Plain washer - 2 off
9 Speedometer case	19 Spring washer - 3 off	29 Nut - 2 off
10 Tachometer case	20 Damper rubber	30 Speedometer drive cable
		31 Tachometer drive cable

15.3 Rear stop lamp switch is fitted to the master cylinder

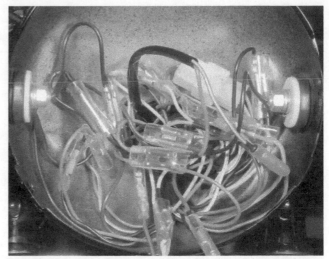

17.1 Wiring is complicated but colour coded to aid identification

13 Speedometer and tachometer head: replacement of bulbs

1   The speedometer and tachometer heads contain four and two bulbs respectively, all of which are rated at 12v 3.4w and are of the bayonet fitting type.

2   The bulbholders are a push fit into the base of the instrument where they are retained by their outer moulded rubber sleeves. Access to the bulbholders can be made after removing the two dome nuts which hold each head in place in its separate outer shell.

14 Warning lamp console: replacement of bulbs

1   In addition to the six bulbs fitted within the instrument heads, a further three warning bulbs are fitted within a warning console to the rear of the instrument assembly. Again, each bulb is rated at 3.4w and is of the bayonet fitting type.

2   Access to the bulbs may be made after removing the console top cover, which is retained from the underside by three screws which pass through the lower cover.

15 Stop lamp switches: location and replacement

1   Two stop lamp switches are fitted to the machine, which work independently of one another, depending on which brake is operated.

2   The front brake switch is fitted to the handlebar lever stock and is of a mechanical push-off type, being operated when the lever is moved. The switch is a push fit in the housing boss.

3   The rear brake stop lamp switch is fitted to the master cylinder and is operated by hydraulic pressure when the foot pedal is depressed. The switch is screwed into the body of the master cylinder. In the event of switch failure, the switch may only be removed after the brake fluid has been drained. After refitting a new switch, the fluid must be replaced and the system bled of all air.

16 Horn: location and examination

1   The horn is suspended from a flexible steel strip bolted immediately below the steering head, between the duplex down tubes of the frame. The flexible strip isolates the horn from vibration.

2   The horn has no external means of adjustment. If it malfunctions, it must be renewed; it is a statutory requirement that the machine must be fitted with a horn in working order.

17 Wiring: layout and examination

1   The wiring harness is colour-coded and will correspond with the accompanying wiring diagram. Where socket connectors are used, they are designed so that reconnection can be made in the correct position only.

2   Visual inspection will show whether there are any breaks or frayed outer coverings which will give rise to short circuits. Another source of trouble may be the snap connectors and sockets, where the connector has not been pushed fully home in the outer housing.

3   Intermittent short circuits can often be traced to a chafed wire that passes through or is close to a metal component such as a frame member. Avoid tight bends in the lead or situations where a lead can become trapped between castings.

18 Ignition and lighting switch

1   The ignition and lighting switch is combined in one unit, bolted to the top fork yoke. It is operated by a key, which cannot be removed when the ignition is switched on.

2   The number stamped on the key will match the number of the steering head lock and that of the lock in the petrol filler cap. A replacement key can be obtained if the number is quoted; if either of the locks or the ignition switch is changed, additional keys will be required.

3   It is not practicable to repair the ignition switch if it malfunctions. It should be renewed with a new switch and key to suit.

See next page for 'Fault diagnosis - electrical system'

19 Fault diagnosis: electrical system

Symptom	Cause	Remedy
Complete electrical failure	Blown fuse	Check wiring and electrical components for short circuit before fitting new 15 amp fuse. Check battery connections, also whether connections show signs of corrosion.
Dim lights, horn inoperative	Discharged battery	Recharge battery with battery charger and check whether alternator is giving correct output (electrical specialist).
Constantly 'blowing' bulbs	Vibration, poor earth connection	Check whether bulb holders are secured correctly. Check earth return or connections to frame.

Wiring diagrams overleaf

1 Front flasher light (R)
2 Horn
3 Pilot lamp
3a Oil
3b Flasher (R)
3c Flasher (L)
4 Headlight
5 Auxiliary light
6 Tachometer
6a Highbeam
6b Neutral
6c Lighting (x 2)
7 Speedometer
7a Lighting (x 2)
7b Sensor
8 Front flasher light (L)
9 Front stop switch
10 Main switch
11 Flasher switch
12 Passing button
13 Horn button
14 Dimmer switch
15 Light switch
16 Engine stop switch
17 Cancelling nut
   (except for Germany)
18 Flasher relay
19 Rear stop switch
20 Fuse box
21 Spark plug
22 Ignition coil
23 Neutral switch
24 AC generator
25 Battery
26 Rectifier
27 Regulator
28 Oil level switch
29 Rear flasher light (R)
30 Tail/Stoplight
31 Rear flasher light (L)

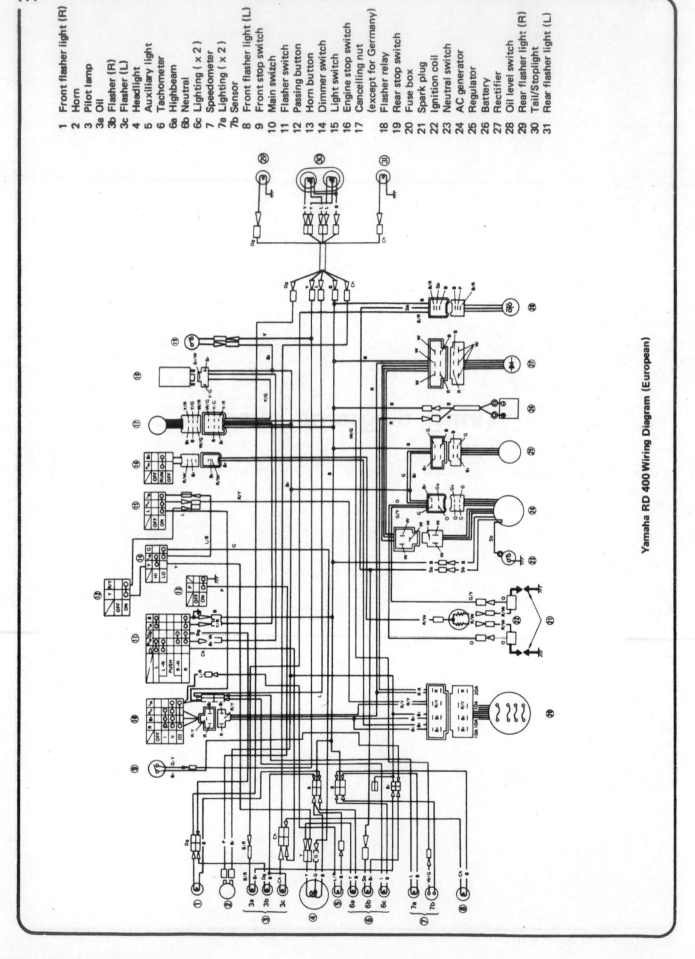

Yamaha RD 400 Wiring Diagram (European)

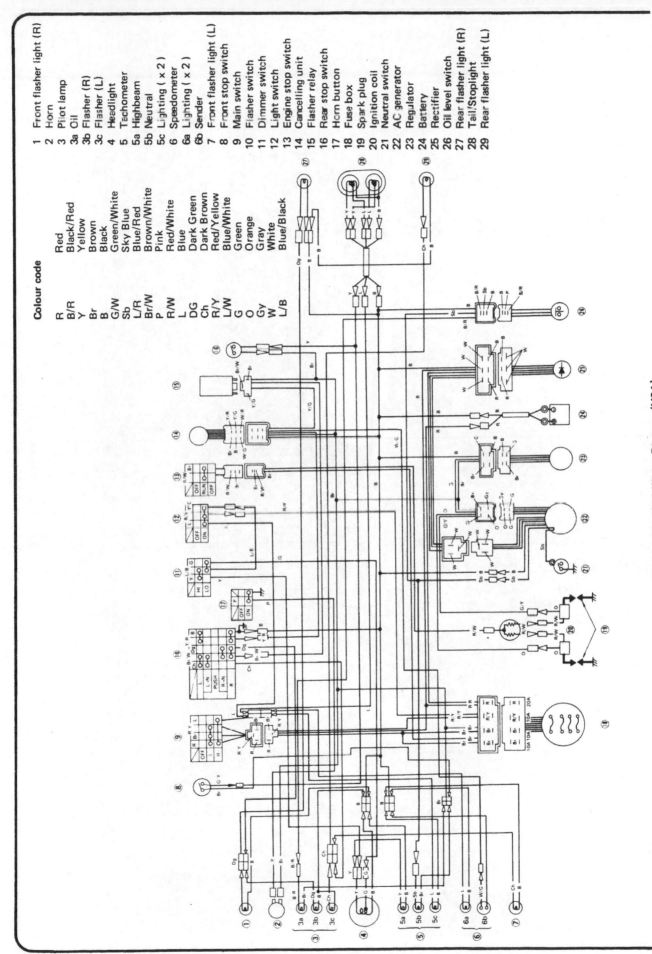

Colour code

R	Red
B/R	Black/Red
Y	Yellow
Br	Brown
B	Black
G/W	Green/White
Sb	Sky Blue
L/R	Blue/Red
Br/W	Brown/White
P	Pink
R/W	Red/White
L	Blue
DG	Dark Green
Ch	Dark Brown
R/Y	Red/Yellow
L/W	Blue/White
G	Green
O	Orange
Gy	Gray
W	White
L/B	Blue/Black

1 Front flasher light (R)
2 Horn
3 Pilot lamp
3a Oil
3b Flasher (R)
3c Flasher (L)
4 Headlight
5 Tachometer
5a Highbeam
5b Neutral
5c Lighting (x 2)
6 Speedometer
6a Lighting (x 2)
6b Sender
7 Front flasher light (L)
8 Front stop switch
9 Main switch
10 Flasher switch
11 Dimmer switch
12 Light switch
13 Engine stop switch
14 Cancelling unit
15 Flasher relay
16 Rear stop switch
17 Horn button
18 Fuse box
19 Spark plug
20 Ignition coil
21 Neutral switch
22 AC generator
23 Regulator
24 Battery
25 Rectifier
26 Oil level switch
27 Rear flasher light (R)
28 Tail/Stoplight
29 Rear flasher light (L)

Yamaha RD 400C Wiring Diagram (USA)

1979 Yamaha RD400E model

Chapter 7 Yamaha RD400E Model

Contents

Specifications

Except where entered below specifications for the RD400E
remain the same as given for earlier models at the beginning of
each Chapter.

Specifications relating to Chapter 2

Carburettor

Type	VM28SS
Main jet	145
Needle jet	0—8
Jet needle	5J6—3
Air jet	0.5

Specifications relating to Chapter 3

Ignition system

Type	Capacitor discharge ignition (CDI)

Flywheel generator

Make	Nippon Denso
Model	032000—054
Output	14V, 13.5A @ 500 rpm minimum

CDI unit

Make	Nippon Denso
Model	070000—0380

Ignition coil

Make	Nippon Denso
Model	0297000—4740

Ignition timing ... ... 2.0 mm (0.08 in) BTDC

Spark plug

Gap	0.7 - 0.8 mm (0.028 - 0.031 in)

Specifications relating to Chapter 4

Front forks

Damping oil capacity	163.5 $\pm$ 4 cc
Damping oil specification	SAE 10W/30 motor oil
Fork oil level from top of fork tube	413 $\pm$ 10 mm (16.26 $\pm$ 0.39 in)
Fork spring free length	394.5 mm (15.53 in)

Specifications relating to Chapter 5

Brakes

Pad size	11 mm (0.43 in)
Service limit	6.5 mm (0.26 in)

Specifications relating to Chapter 6

Flywheel generator

Make	...	...	...	...	...	...	...	...	Nippon Denso
Model	...	...	...	...	...	...	...	...	032000–054
Output	...	...	...	...	...	...	...	...	14V, 13.5A @ 500 rpm minimum

Rectifier/regulator

Make	...	...	...	...	...	...	...	...	Shindengen
Type	...	...	...	...	...	...	...	...	SH235

Bulbs

Headlamp	...	...	...	...	...	...	...	...	35/35 Watt

1 General description

The most significant change made to the Yamaha RD400E is the adoption of electronic ignition of the capacitor discharge ignition type. With this system the contact breaker is dispensed with as a means of interrupting the primary ignition circuit and thus causing a high voltage to be induced in the ignition coil. Instead, a pulser unit and transistor switching circuit is employed to fulfil the same function. Discarding the contact breaker, which is a mechanical unit subject to wear, enables greater accuracy in ignition timing to be maintained and reduces the amount of maintenance the ignition requires. In addition the performance of the ignition system is reflected in the performance of the machine.

Although hydraulically operated disc brakes are still employed on the RD400E, the brake calipers have been changed from the two-piston fixed type to the single piston floating type. As a result the maintenance and service procedures are somewhat changed.

A number of minor modifications which do not substantially affect servicing or overhaul have been made to the frame and engine. These changes are listed in Section 8 of this Chapter. Where specificational changes have been made these are given in the specifications at the beginning of the Chapter.

2 Ignition timing: checking and adjusting

1 Because no mechanical contact breaker is fitted to the ignition system no alteration of the ignition timing should be experienced during normal service of the machine, and therefore maintenance can be reduced to a minimum. If, however, the performance of the machine deteriorates or its general health is suspect the ignition timing should be checked and if necessary adjusted.

2 To check the ignition timing a dial test indicator (DTI) and adapter is required which is screwed into the spark plug hole and enables the position of the piston to be determined. Remove the flywheel generator cover, and the spark plug from the left-hand cylinder. Install the DTI and set it so that it reads zero at **exactly** top dead centre. Rotate the engine clockwise about 45° and then slowly turn the engine anti-clockwise until the DTI indicates that the piston is **exactly** 2.0 mm (0.08 in) from TDC. Observe the timing mark '1F' on the periphery of the generator flywheel. This mark should be exactly in line with the index mark on the pulser coil which may be found in the 2 o'clock position on the generator stator plate. If the marks are in alignment the ignition timing is correct.

3 To adjust the timing slacken the three stator retaining bolts which pass through elongated holes disposed equally around the periphery of the stator plate. The plate should be rotated to bring the '1F' mark into alignment with the index mark. This may be accomplished by using a screwdriver as a lever, placed against the grooves provided in the plate in the 7 o'clock position. When the alignment is correct tighten the three bolts.

4 An alternative method of ignition timing checking may be made with the engine running, using a stroboscopic lamp. The lamp beam should be aimed at the index mark with the engine running at about 1000 rpm. The engine must be stopped for adjustment to be carried out.

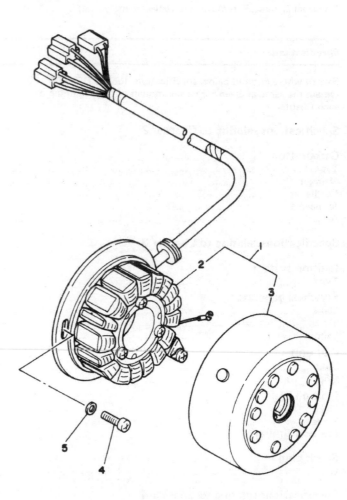

Fig. 7.1 Flywheel generator

1 *Generator assembly*
2 *Stator coil assembly*
3 *Rotor*
4 *Adjusting screw – 2 off*
5 *Plain washer – 2 off*

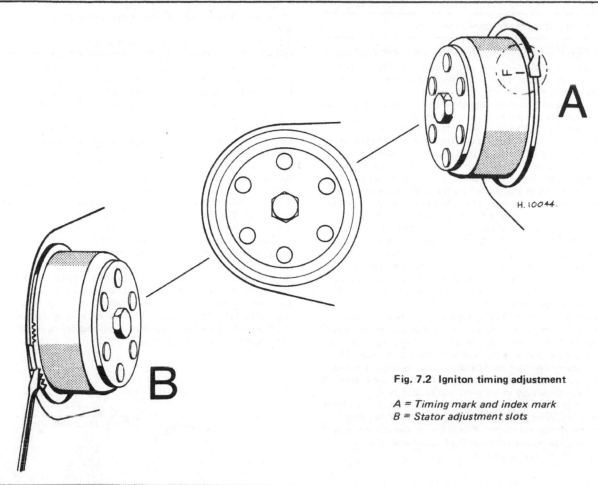

H. 10044.

Fig. 7.2 Igniton timing adjustment

A = Timing mark and index mark
B = Stator adjustment slots

3 CDI system: testing

1   If the performance of the ignition system is suspect the various components should be checked to eliminate the faulty item. This will require the use of a multi-meter set to the resistance measuring position or a separate ohmmeter.
2   To check the condition of the pulser coil and ignition source coil within the flywheel generator disconnect the five wires leading from the stator at their individual block connectors. Test the resistance across the three pairs of wires as follows:

Charging coil
   Red to brown (high speed)   *5.1 ohms ± 10% at 20°C (68°F)*
   Brown to black (low speed)   *271 ohms ± 10% at 20°C (68°F)*
Pulser coil
   White/red to black   *87 ohms ± 10% at 20°C (68°F)*

The resistance varies somewhat with temperature, but if a reading taken is well outside that given it may be assumed that the coil has malfunctioned.
3   A visual inspection of the coils may be made after removing the flywheel rotor. This is accomplished as described in the following section.
4   If the generator coils are found to be in good condition the single ignition (secondary) coil should be checked, once again using a multi-meter to measure the resistance. Test the coil after disconnecting the HT leads from the suppressor caps and after separating the low tension lead block connector. If the coil test resistances are not as given below the coil must be renewed.

   Primary coil   *0.33 ohms ± 20% at 20°C (68°F)*
   Secondary coil   *3.5 Kohms ± 30% at 20°C (68°F)*

5   Should it be found that the performance of both the flywheel generator coils and the ignition coil is satisfactory, there is evidence that the CDI unit has failed. Unfortunately the testing of the CDI circuits requires special equipment and it is recommended, therefore, that the unit be returned to a Yamaha Service Agent or an auto-electrician for testing.

4 Flywheel generator: removal and replacement

1   Remove the left-hand crankcase cover so that access may be made to the flywheel generator rotor. The rotor is secured on the crankshaft end by a central nut. To prevent the rotor turning when loosening the nut place the gearbox in top gear and apply firmly the rear brake. The rotor is positioned on the tapered end of the crankshaft and located by a Woodruff key. Because of this the rotor will be very tight and will require pulling from position. To aid this the rotor centre is internally threaded to accept a small extractor (Yamaha tool number 90890−01189). Removal of the rotor without the extractor, by using levers, is not recommended as it will almost certainly lead to damage to the casings or rotor.
2   To displace the rotor, screw the extractor into the rotor centre until it is fully home and then tighten down the centre bolt until it abuts against the crankshaft end. Further tightening of the screw should release the rotor from the taper. If the rotor is reluctant to move do not overtighten the centre screw. A few sharp taps on the screw head with a hammer will usually produce the desired results.
3   Stator removal is straightforward, requiring removal of the three stator plate adjustment/retaining bolts. To aid retiming the ignition, mark the plate in relation to the casing so that it can be refitted in the same position.
4   The flywheel generator may be refitted by reversing the dismantling sequence. The ignition timing should be checked as a matter of course as described in Section 2.

5 Front and rear disc brakes: general description

1　The RD400E models have been fitted with a different type of caliper unit on both front and rear disc brakes. The new units are of the floating caliper type and a new procedure is required when dealing with them.

2　Before attempting to start any work on the caliper unit check which type has been fitted to your machine. Both types work on broadly the same principle, the fixed caliper having two pistons, the floating caliper having only one. When the front brake lever is depressed, it causes the master piston to move within the master cylinder, to which it is linked. As the piston moves in the cylinder, it traps the brake fluid causing a pressure build-up which is transmitted to the caliper via the brake hose and pipe that forms the connecting link. The pressure in the caliper cylinders causes the brake pads of a fixed type caliper to move in their respective housings and bear against the disc, the friction between the two pads and the disc provides the brake action. As the brake lever is released, the pressure falls and the piston seals which had been distorted when the pistons moved towards the disc, now revert to their normal shape and position pulling the pistons back with them. The pads no longer bear on the disc and the wheel is again free to revolve.

3　If the caliper is of the floating type it has only one piston. When the single piston bears upon the disc the caliper unit slides fractionally to the right to bring the second fixed caliper into use. The caliper is mounted on a pivot to allow this to happen. This provides adequate pressure on the disc surfaces without the added complexities of a double piston caliper. The hydraulic pressure is supplied in the same way as for the fixed caliper type.

4　The leverage of the brake lever is such that it produces a force at the master cylinder piston approximately four times that applied to the brake lever itself. This is one of the reasons why the hydraulic braking system is more efficient than the older more conventional drum brake type.

6 Front disc brake: removing and replacing the disc and pads

1　The actual brake disc, which is bolted to the right-hand side of the machine, rarely requires attention. Check the disc for signs of warping and wear. Warping of the disc may have resulted from crash damage and should not exceed 0.15 mm at any point. The disc itself must not be allowed to wear below the limit thickness of 6.5 mm. If these figures are exceeded in any case, the disc must be renewed.

2　To remove the disc from the wheel, it is first necessary to detach the front wheel from the machine. Place the machine on the centre stand so that the front wheel is raised clear of the ground. It may be necessary to place thin wooden blocks between the stand feet and the ground to assist with this. Detach the speedometer cable after unscrewing the knurled retaining ring. Remove the split pin and castellated nut from the end of the wheel spindle. After slackening the clamp on the opposite fork leg, withdraw the spindle, using a suitable tommy bar. Remove the wheel complete, taking care not to damage the disc or caliper. A small wooden wedge can be inserted in the caliper between the brake pads to prevent them being expelled if the brake lever should be operated inadvertently. There is in any case a tendency to 'creep' in hydraulic systems, producing the same results.

3　The disc itself is attached directly to the hub. The four mounting bolts are locked into position by two double tab washers, which must always be renewed on reassembly. The disc surfaces may now be examined for scoring, which is usually caused by particles of grit becoming embedded in the friction material of the pads. If excessive, the disc should be renewed as deep scoring is detrimental to braking efficiency. Reassembly is a direct reversal of the dismantling procedure. Ensure that each component is thoroughly cleaned, especially mating surfaces to obviate the risk of mal-alignment. Remember to bend up the ears

of the double tab washers against the bolt heads to secure the bolts.

4　Pad removal may be accomplished without detaching the caliper unit from the fork leg if the front wheel has been removed. If wheel removal has not taken place the caliper unit must be removed. Disconnection of the hydraulic hose is not, however, required. The caliper may be detached by displacing the plastic end plug and withdrawing the mounting bolt upon which the caliper unit pivots.

5　The pads are secured by a pin which is located by a coil spring. Displace the spring so that the pin can be withdrawn, and lift out the pads. Examine each pad for wear. If it is found that either pad has worn down to the wear limit groove on the pad periphery or has an overall thickness of less than 6.5 mm (0.26 in) both pads should be renewed.

6　Reassembly may be carried out by reversing the dismantling procedure. The manufacturer recommends that the pad retaining pin and its locating spring is renewed when the pads are renewed as a safety precaution. If new pads are fitted some difficulty may be encountered when placing the assembled caliper unit over the disc, due to the reduction in the gap between the pad faces. If this proves to be the case the moving pad may be pushed inwards against the piston so that the piston moves back and the necessary clearance is obtained. The caliper mounting bolt should be lubricated with grease before it is inserted and tightened.

7　The brakes **must** be checked for satisfactory operation before the machine is taken on the road.

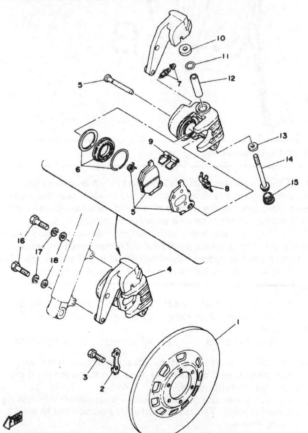

Fig. 7.3　Front disc brake caliper

1	Disc	10	Washer
2	Double tab washer – 3 off	11	'O'-ring
3	Bolt – 6 off	12	Sleeve
4	Front caliper assembly	13	Washer
5	Brake pad set	14	Bolt
6	Caliper seal assembly	15	Swing bolt cap
7	Bleed screw and cap	16	Bolt – 2 off
8	Spring	17	Spring washer – 2 off
9	Retainer	18	Washer – 2 off

7 Rear disc brake: removing and replacing the disc and pads

1   The brake caliper unit fitted to the rear of the machine is similar to that used on the front, and as such servicing procedures are fundamentally the same as those given in the previous Section.

8 Other minor modifications

1   The kickstart mechanism is now of the bendix type, rather than the previously employed ratchet type. As can be seen from Fig. 7.4 the various components are mounted in a similar fashion on the kickstart shaft and dismantling and assembly is straightforward.

2   The carburettors are now fitted with an inspection window in the main body in place of the inspection plug which was fitted formerly. In addition the jet sizes and specifications have been altered. Refer to the specifications at the beginning of this Chapter for details. A balance tube interlinking the carburettors assists in maintaining carburettor synchronisation.

3   The two separate ignition coils fitted to earlier models have been replaced by a single coil to match the CDI system.

4   The voltage regulator and rectifier are combined in a single unit, rather than being separate units.

5   The position of the footrests and handlebars have been changed slightly and the front fork travel has been increased. These changes have been made to improve rider comfort.

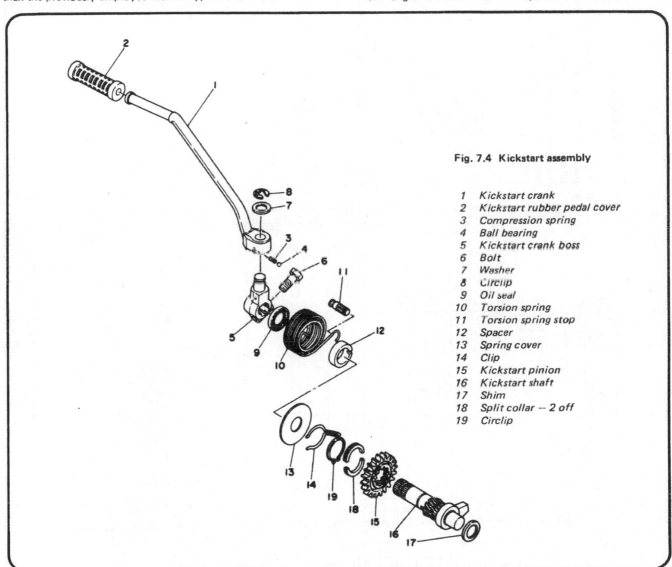

Fig. 7.4 Kickstart assembly

1   Kickstart crank
2   Kickstart rubber pedal cover
3   Compression spring
4   Ball bearing
5   Kickstart crank boss
6   Bolt
7   Washer
8   Circlip
9   Oil seal
10 Torsion spring
11 Torsion spring stop
12 Spacer
13 Spring cover
14 Clip
15 Kickstart pinion
16 Kickstart shaft
17 Shim
18 Split collar -- 2 off
19 Circlip

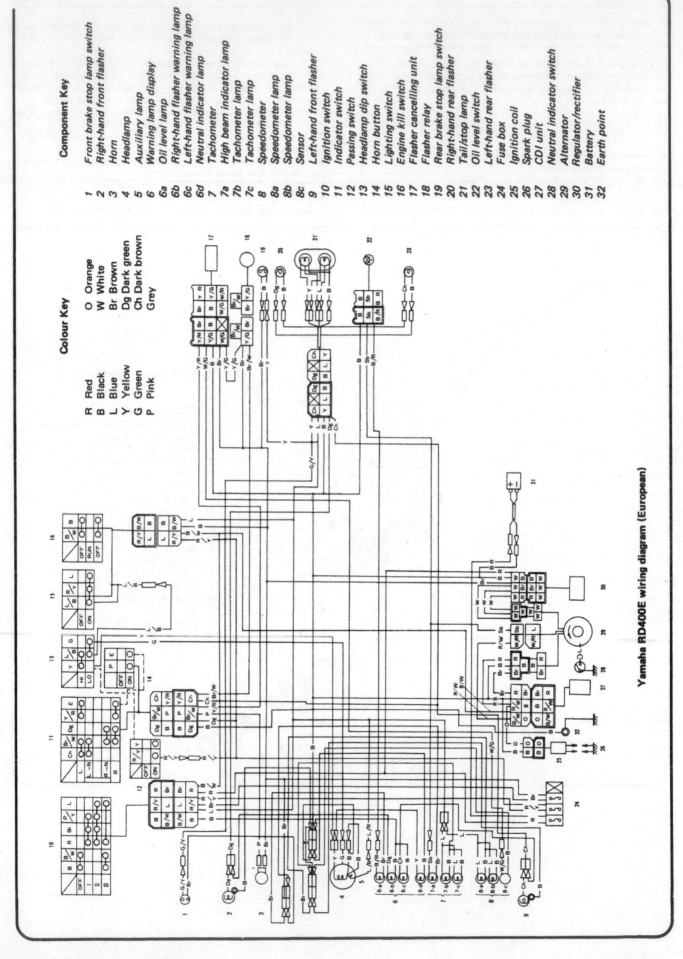

Component Key

1 Front brake stop lamp switch
2 Right-hand front flasher
3 Horn
4 Headlamp
5 Auxiliary lamp
6 Warning lamp display
6a Oil level lamp
6b Right-hand flasher warning lamp
6c Left-hand flasher warning lamp
6d Neutral indicator lamp
7 Tachometer
7a High beam indicator lamp
7b Tachometer lamp
7c Tachometer lamp
8 Speedometer
8a Speedometer lamp
8b Speedometer lamp
8c Sensor
9 Left-hand front flasher
10 Ignition switch
11 Indicator switch
12 Passing switch
13 Headlamp dip switch
14 Horn button
15 Engine kill switch
16 Lighting switch
17 Flasher cancelling unit
18 Flasher relay
19 Rear brake stop lamp switch
20 Right-hand rear flasher
21 Tail/stop lamp
22 Oil level switch
23 Left-hand rear flasher
24 Fuse box
25 Ignition coil
26 Spark plug
27 CDI unit
28 Neutral indicator switch
29 Alternator
30 Regulator/rectifier
31 Battery
32 Earth point

Colour Key

R	Red	O	Orange
B	Black	W	White
L	Blue	Br	Brown
Y	Yellow	Dg	Dark green
G	Green	Ch	Dark brown
P	Pink		Grey

Yamaha RD400E wiring diagram (European)

Metric conversion tables

Inches	Decimals	Millimetres	Millimetres to Inches		Inches to Millimetres	
			mm	Inches	Inches	mm
1/64	0.015625	0.3969	0.01	0.00039	0.001	0.0254
1/32	0.03125	0.7937	0.02	0.00079	0.002	0.0508
3/64	0.046875	1.1906	0.03	0.00118	0.003	0.0762
1/16	0.0625	1.5875	0.04	0.00157	0.004	0.1016
5/64	0.078125	1.9844	0.05	0.00197	0.005	0.1270
3/32	0.09375	2.3812	0.06	0.00236	0.006	0.1524
7/64	0.109375	2.7781	0.07	0.00276	0.007	0.1778
1/8	0.125	3.1750	0.08	0.00315	0.008	0.2032
9/64	0.140625	3.5719	0.09	0.00354	0.009	0.2286
5/32	0.15625	3.9687	0.1	0.00394	0.01	0.254
11/64	0.171875	4.3656	0.2	0.00787	0.02	0.508
3/16	0.1875	4.7625	0.3	0.1181	0.03	0.762
13/64	0.203125	5.1594	0.4	0.01575	0.04	1.016
7/32	0.21875	5.5562	0.5	0.01969	0.05	1.270
15/64	0.234275	5.9531	0.6	0.02362	0.06	1.524
1/4	0.25	6.3500	0.7	0.02756	0.07	1.778
17/64	0.265625	6.7469	0.8	0.3150	0.08	2.032
9/32	0.28125	7.1437	0.9	0.03543	0.09	2.286
19/64	0.296875	7.5406	1	0.03937	0.1	2.54
5/16	0.3125	7.9375	2	0.07874	0.2	5.08
21/64	0.328125	8.3344	3	0.11811	0.3	7.62
11/32	0.34375	8.7312	4	0.15748	0.4	10.16
23/64	0.359375	9.1281	5	0.19685	0.5	12.70
3/8	0.375	9.5250	6	0.23622	0.6	15.24
25/64	0.390625	9.9219	7	0.27559	0.7	17.78
13/32	0.40625	10.3187	8	0.31496	0.8	20.32
27/64	0.421875	10.7156	9	0.35433	0.9	22.86
7/16	0.4375	11.1125	10	0.39270	1	25.4
29/64	0.453125	11.5094	11	0.43307	2	50.8
15/32	0.46875	11.9062	12	0.47244	3	76.2
31/64	0.484375	12.3031	13	0.51181	4	101.6
1/2	0.5	12.7000	14	0.55118	5	127.0
33/64	0.515625	13.0969	15	0.59055	6	152.4
17/32	0.53125	13.4937	16	0.62992	7	177.8
35/64	0.546875	13.8906	17	0.66929	8	203.2
9/16	0.5625	14.2875	18	0.70866	9	228.6
37/64	0.578125	14.6844	19	0.74803	10	254.0
19/32	0.59375	15.0812	20	0.78740	11	279.4
39/64	0.609375	15.4781	21	0.82677	12	304.8
5/8	0.625	15.8750	22	0.86614	13	330.2
41/64	0.640625	16.2719	23	0.90551	14	355.6
21/32	0.65625	16.6687	24	0.94488	15	381.0
43/64	0.671875	17.0656	25	0.98425	16	406.4
11/16	0.6875	17.4625	26	1.02362	17	431.8
45/64	0.703125	17.8594	27	1.06299	18	457.2
23/32	0.71875	18.2562	28	1.10236	19	482.6
47/64	0.734375	18.6531	29	1.14173	20	508.0
3/4	0.75	19.0500	30	1.18110	21	533.4
49/64	0.765625	19.4469	31	1.22047	22	558.8
25/32	0.78125	19.8437	32	1.25984	23	584.2
51/64	0.796875	20.2406	33	1.29921	24	609.6
13/16	0.8125	20.6375	34	1.33858	25	635.0
53/64	0.828125	21.0344	35	1.37795	26	660.4
27/32	0.84375	21.4312	36	1.41732	27	685.8
55/64	0.859375	21.8281	37	1.4567	28	711.2
7/8	0.875	22.2250	38	1.4961	29	736.6
57/64	0.890625	22.6219	39	1.5354	30	762.0
29/32	0.90625	23.0187	40	1.5748	31	787.4
59/64	0.921875	23.4156	41	1.6142	32	812.8
15/16	0.9375	23.8125	42	1.6535	33	838.2
61/64	0.953125	24.2094	43	1.6929	34	863.6
31/32	0.96875	24.6062	44	1.7323	35	889.0
63/64	0.984375	25.0031	45	1.7717	46	914.4

English/American terminology

Because this book has been written in England, British English component names, phrases and spellings have been used throughout. American English usage is quite often different and whereas normally no confusion should occur, a list of equivalent terminology is given below.

English	American	English	American
Air filter	Air cleaner	Number plate	License plate
Alignment (headlamp)	Aim	Output or layshaft	Countershaft
Allen screw/key	Socket screw/wrench	Panniers	Side cases
Anticlockwise	Counterclockwise	Paraffin	Kerosene
Bottom/top gear	Low/high gear	Petrol	Gasoline
Bottom/top yoke	Bottom/top triple clamp	Petrol/fuel tank	Gas tank
Bush	Bushing	Pinking	Pinging
Carburettor	Carburetor	Rear suspension unit	Rear shock absorber
Catch	Latch	Rocker cover	Valve cover
Circlip	Snap ring	Selector	Shifter
Clutch drum	Clutch housing	Self-locking pliers	Vise-grips
Dip switch	Dimmer switch	Side or parking lamp	Parking or auxiliary light
Disulphide	Disulfide	Side or prop stand	Kick stand
Dynamo	DC generator	Silencer	Muffler
Earth	Ground	Spanner	Wrench
End float	End play	Split pin	Cotter pin
Engineer's blue	Machinist's dye	Stanchion	Tube
Exhaust pipe	Header	Sulphuric	Sulfuric
Fault diagnosis	Trouble shooting	Sump	Oil pan
Float chamber	Float bowl	Swinging arm	Swingarm
Footrest	Footpeg	Tab washer	Lock washer
Fuel/petrol tap	Petcock	Top box	Trunk
Gaiter	Boot	Torch	Flashlight
Gearbox	Transmission	Two/four stroke	Two/four cycle
Gearchange	Shift	Tyre	Tire
Gudgeon pin	Wrist/piston pin	Valve collar	Valve retainer
Indicator	Turn signal	Valve collets	Valve cotters
Inlet	Intake	Vice	Vise
Input shaft or mainshaft	Mainshaft	Wheel spindle	Axle
Kickstart	Kickstarter	White spirit	Stoddard solvent
Lower leg	Slider	Windscreen	Windshield
Mudguard	Fender		

Index

Conversion Factors

Length (distance)

Inches (in)	x 25.4	= Millimetres (mm)	x 0.0394	= Inches (in)
Feet (ft)	x 0.305	= Metres (m)	x 3.281	= Feet (ft)
Miles	x 1.609	= Kilometres (km)	x 0.621	= Miles

Volume (capacity)

Cubic inches (cu in; in³)	x 16.387	= Cubic centimetres (cc; cm³)	x 0.061	= Cubic inches (cu in; in³)
Imperial pints (Imp pt)	x 0.568	= Litres (l)	x 1.76	= Imperial pints (Imp pt)
Imperial quarts (Imp qt)	x 1.137	= Litres (l)	x 0.88	= Imperial quarts (Imp qt)
Imperial quarts (Imp qt)	x 1.201	= US quarts (US qt)	x 0.833	= Imperial quarts (Imp qt)
US quarts (US qt)	x 0.946	= Litres (l)	x 1.057	= US quarts (US qt)
Imperial gallons (Imp gal)	x 4.546	= Litres (l)	x 0.22	= Imperial gallons (Imp gal)
Imperial gallons (Imp gal)	x 1.201	= US gallons (US gal)	x 0.833	= Imperial gallons (Imp gal)
US gallons (US gal)	x 3.785	= Litres (l)	x 0.264	= US gallons (US gal)

Mass (weight)

Ounces (oz)	x 28.35	= Grams (g)	x 0.035	= Ounces (oz)
Pounds (lb)	x 0.454	= Kilograms (kg)	x 2.205	= Pounds (lb)

Force

Ounces-force (ozf; oz)	x 0.278	= Newtons (N)	x 3.6	= Ounces-force (ozf; oz)
Pounds-force (lbf; lb)	x 4.448	= Newtons (N)	x 0.225	= Pounds-force (lbf; lb)
Newtons (N)	x 0.1	= Kilograms-force (kgf; kg)	x 9.81	= Newtons (N)

Pressure

Pounds-force per square inch (psi; lbf/in²; lb/in²)	x 0.070	= Kilograms-force per square centimetre (kgf/cm²; kg/cm²)	x 14.223	= Pounds-force per square inch (psi; lbf/in²; lb/in²)
Pounds-force per square inch (psi; lbf/in²; lb/in²)	x 0.068	= Atmospheres (atm)	x 14.696	= Pounds-force per square inch (psi; lbf/in²; lb/in²)
Pounds-force per square inch (psi; lbf/in²; lb/in²)	x 0.069	= Bars	x 14.5	= Pounds-force per square inch (psi; lbf/in²; lb/in²)
Pounds-force per square inch (psi; lbf/in²; lb/in²)	x 6.895	= Kilopascals (kPa)	x 0.145	= Pounds-force per square inch (psi; lbf/in²; lb/in²)
Kilopascals (kPa)	x 0.01	= Kilograms-force per square centimetre (kgf/cm²; kg/cm²)	x 98.1	= Kilopascals (kPa)
Millibar (mbar)	x 100	= Pascals (Pa)	x 0.01	= Millibar (mbar)
Millibar (mbar)	x 0.0145	= Pounds-force per square inch (psi; lbf/in²; lb/in²)	x 68.947	= Millibar (mbar)
Millibar (mbar)	x 0.75	= Millimetres of mercury (mmHg)	x 1.333	= Millibar (mbar)
Millibar (mbar)	x 0.401	= Inches of water (inH₂O)	x 2.491	= Millibar (mbar)
Millimetres of mercury (mmHg)	x 0.535	= Inches of water (inH₂O)	x 1.868	= Millimetres of mercury (mmHg)
Inches of water (inH₂O)	x 0.036	= Pounds-force per square inch (psi; lbf/in²; lb/in²)	x 27.68	= Inches of water (inH₂O)

Torque (moment of force)

Pounds-force inches (lbf in; lb in)	x 1.152	= Kilograms-force centimetre (kgf cm; kg cm)	x 0.868	= Pounds-force inches (lbf in; lb in)
Pounds-force inches (lbf in; lb in)	x 0.113	= Newton metres (Nm)	x 8.85	= Pounds-force inches (lbf in; lb in)
Pounds-force inches (lbf in; lb in)	x 0.083	= Pounds-force feet (lbf ft; lb ft)	x 12	= Pounds-force inches (lbf in; lb in)
Pounds-force feet (lbf ft; lb ft)	x 0.138	= Kilograms-force metres (kgf m; kg m)	x 7.233	= Pounds-force feet (lbf ft; lb ft)
Pounds-force feet (lbf ft; lb ft)	x 1.356	= Newton metres (Nm)	x 0.738	= Pounds-force feet (lbf ft; lb ft)
Newton metres (Nm)	x 0.102	= Kilograms-force metres (kgf m; kg m)	x 9.804	= Newton metres (Nm)

Power

Horsepower (hp)	x 745.7	= Watts (W)	x 0.0013	= Horsepower (hp)

Velocity (speed)

Miles per hour (miles/hr; mph)	x 1.609	= Kilometres per hour (km/hr; kph)	x 0.621	= Miles per hour (miles/hr; mph)

Fuel consumption*

Miles per gallon (mpg)	x 0.354	= Kilometres per litre (km/l)	x 2.825	= Miles per gallon (mpg)

Temperature

Degrees Fahrenheit = (°C x 1.8) + 32   Degrees Celsius (Degrees Centigrade; °C) = (°F - 32) x 0.56

It is common practice to convert from miles per gallon (mpg) to litres/100 kilometres (l/100km), where mpg x l/100 km = 282

Preserving Our Motoring Heritage

<
The Model J Duesenberg Derham Tourster. Only eight of these magnicent cars were ever built – this is the only example to be found outside the United States of America

Almost every car you've ever loved, loathed or desired is gathered under one roof at the Haynes Motor Museum. Over 300 immaculately presented cars and motorbikes represent every aspect of our motoring heritage, from elegant reminders of bygone days, such as the superb Model J Duesenberg to curiosities like the bug-eyed BMW Isetta. There are also many old friends and flames. Perhaps you remember the 1959 Ford Popular that you did your courting in? The magnificent 'Red Collection' is a spectacle of classic sports cars including AC, Alfa Romeo, Austin Healey, Ferrari, Lamborghini, Maserati, MG, Riley, Porsche and Triumph.

A Perfect Day Out

Each and every vehicle at the Haynes Motor Museum has played its part in the history and culture of Motoring. Today, they make a wonderful spectacle and a great day out for all the family. Bring the kids, bring Mum and Dad, but above all bring your camera to capture those golden memories for ever. You will also find an impressive array of motoring memorabilia, a comfortable 70 seat video cinema and one of the most extensive transport book shops in Britain. The Pit Stop Cafe serves everything from a cup of tea to wholesome, home-made meals or, if you prefer, you can enjoy the large picnic area nestled in the beautiful rural surroundings of Somerset.

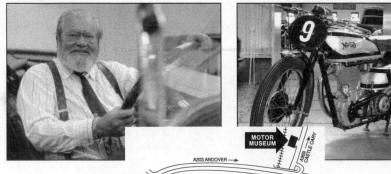

>
John Haynes O.B.E., Founder and Chairman of the museum at the wheel of a Haynes Light 12.

<
The 1936 490cc sohc-engined International Norton – well known for its racing success

The Museum is situated on the A359 Yeovil to Frome road at Sparkford, just off the A303 in Somerset. It is about 40 miles south of Bristol, and 25 minutes drive from the M5 intersection at Taunton.
Open 9.30am - 5.30pm (10.00am - 4.00pm Winter) 7 days a week, *except Christmas Day, Boxing Day and New Years Day*
Special rates available for schools, coach parties and outings Charitable Trust No. 292048